P9-CFX-702

# BRAND sense

***Build Powerful Brands through Touch, Taste, Smell, Sight, and Sound***

**Martin Lindstrom**

*Foreword by Philip Kotler*

Free Press
NEW YORK • LONDON • TORONTO • SYDNEY

*f*P
FREE PRESS
A Division of Simon & Schuster, Inc.
1230 Avenue of the Americas
New York, NY 10020



For information about special discounts for bulk purchases,
please contact Simon & Schuster Special Sales:
1-800-456-6798 or business@simonandschuster.com

TEXT DESIGNED BY PAUL DIPPOLITO

All illustrations were created by Martin Lindstrom.

Manufactured in the United States of America

10 9 8 7 6

Library of Congress Cataloging-in-Publication Data
Lindstrom, Martin
Brand sense: build powerful brands through touch, taste,
smell, sight, and sound / Martin Lindstrom; foreword by Philip Kotler.
p. cm.
Includes bibliographical references and index.
1. Brand name products. 2. Business names. 3. Advertising—Brand name
products. 4. Advertising—Psychological aspects. 5. Senses and sensation.
I. Title.
HD69.B7 L548 2005
658.8'27—dc22 2004056438

ISBN-13: 978-0-7432-6784-7
ISBN-10: 0-7432-6784-2

*For Dorit, Tore, Vibeke, and Allan*
*You are the words in my life*

*Tell me and I'll forget,*
*Show me and I might remember,*
*Involve me and I'll understand.*

—BENJAMIN FRANKLIN

# Contents

# Foreword

*Philip Kotler*

Marketing isn't working today. New products are failing at a disastrous rate. Most advertising campaigns do not register anything distinctive in the customer's mind. Direct mail barely achieves a 1 percent response rate. Most products come across as interchangeable commodities rather than powerful brands.

Yes, there are still powerful brands: Coca-Cola, Harley-Davidson, Apple Computer, Singapore Airlines, BMW. These corporations have learned how to make their brands live in the customers' minds. A brand, of course, must at least deliver a distinctive benefit. No amount of dressing up will make up for this lack. All of the aforementioned brands deliver a distinctive benefit.

But distinctive brands require something more. They have to be powered up to deliver a full sensory and emotional experience. It is not enough to present a product or service visually in an ad. It pays to attach a sound, such as music or powerful words and symbols. The combination of visual and audio stimuli delivers a 2 + 2 = 5 impact. It pays even more to trigger other sensory channels—taste, touch, smell—to enhance the total impact. This is Martin Lindstrom's basic message, and he illustrates it beautifully through numerous cases with compelling arguments.

Most companies take the easy way out to market their brands. They buy a lot of expensive advertising and make clichéd claims. The companies Martin describes are much more creative. One of the main reasons to read this book is that it contains a treasury of ideas for bringing new life to your own brands.

# **BRAND** sense

CHAPTER 1

# A Cottage Industry Turns Professional

**JANUARY 14, 2004 WAS A LANDMARK** in the life of Sydney-born teenager Wilhelm Andries Petrus Booyse. It was a day that passed unnoticed by most, but it proved to be the highlight of Will's life. He lay face-down on a firm table and submitted his neck to the pain of the plastic surgeon's laser.

The doctor worked slowly and diligently, carefully obliterating the tattooed bar code with the letters G-U-C-C-I neatly etched underneath. The beam followed the shape once so carefully duplicated from the Gucci Corporation's printed guidelines. Bit by bit the tattoo was removed. The process was painful, but it marked the end of Will's obsession with the Gucci brand—an obsession he had taken to the outermost limits. Gucci had become more than a brand. It was, in Will's words, "My one and only religion."

I first met Will in May 1999, when his Gucci tattoo was brand spanking new. He had, he believed, formed a lifelong relationship with the brand. Lifelong turned out to be only five years. In that time, the brand was no longer just a brand. For Will it had become a "person" whom he could relate to, admire, and be supported by. This relationship gave him the energy he required to get up each day and go to school. It gave him a sense of his own identity.

He talked about Gucci as a family member, not as an expensive fashion product. He could expound at great length about the designs, the colors, the feel of the fabrics, the texture of the leather, and the distinct smell of the perfumed Gucci environment.

By the time Will removed the Gucci bar code from his neck, he had the sense that the brand was losing its grip. What was once perceived as the ultimate brand, made in heaven, seemed to be slipping. Will was not alone in his perception. Gucci's lack of innovation and dated advertising campaigns suffered a final blow when Tom Ford, Gucci's head designer, unceremoniously decided to go his separate way.

Additionally Will had found another path to follow. The Australian Navy beckoned, offering him another sense of family and identity. A lot of his newfound mates sported tattoos as well, but they generally chose the name of the vessel that they called home for six months each year.

Will summed up his experience with Gucci: "The admiration I had for the Gucci brand was stronger than any other person I knew. For me, Gucci was more than a brand—it was my personal companion. When I entered a Gucci store, I felt like I was in heaven. Everything about the place made me feel at home. The atmosphere of luxury, the lighting, the design, and the music. I suppose the status that this gave me amongst my friends made me an exclusive member of this distinct brand community. In the time that I wore the Gucci tattoo, people approached me constantly and made me feel the center of the universe.

"I don't know what happened, but one day I woke up and the magic was gone. Gucci failed to excite me as it always had. The only thing that remained was the tattoo I'd had done so willingly five years before. So despite the pain of removal, I felt it had to go."

As frightening, shocking, or intriguing as Will's story might sound, my meeting with him sparked the first sensory branding research project ever conducted. It was a five-year mission which involved hundreds of researchers and thousands of consumers across four continents. We sought to understand the rationale behind behavior like Will's.

Will was a living breathing example of what marketers ultimately aspire to when they create a brand. He also was a perfect test subject in our quest to understand the dynamics of strong branding used correctly—and incorrectly. What was it that made a kid base his life on a brand? What components of the brand formed such a magnetic connection? And then, at what point did the brand fail? How did obsessive belief turn into disappointment?

We went out and asked all kinds of questions of people who have particular affinities for various brands. They willingly, and kindly, shared their passions. This invaluable information led me to conclude that if branding wishes to survive another century it will need to change track. More communication in an already overcrowded world simply won't do it. A new vision with an emotional basis is required.

I realized that a brand would have to become a sensory experience that extends beyond the traditional paradigm, which primarily addresses sight and sound.

Another aspect of the new branding that I gleaned from Will is that a brand should create a following similar to the obsessive commitment of sports fans or even, in certain respects, to the faith of a religious community. The bond it forms is the social glue that links and unites generations of people.

Religion, however, is only one side of the story for the next generation of branding. In order to have a viable future, brands will have to incorporate a brand platform that fully integrates the five senses. This sensory platform will reveal the very belief—or significant following—necessary to create a brand philosophy. Without taking comparisons to religion too far, we can see its relevance for some points of sensory branding.

## *Branding: The Next Generation*

The concept of branding is already undergoing dramatic changes. New technologies have allowed us to go beyond mass production and to mass customize brands. Currently brand manufacturers own their brands. This is changing. In the future brands will increasingly be owned by the consumer. The first signs of this shift appeared in the late 1990s. I documented this phenomenon in *BRANDchild* and named it MSP—Me Selling Proposition.

In the 1950s branding belonged to the USP—the Unique Selling Proposition. This ensured that the physical product, rather than the brand, was the core differential. By the 1960s we began seeing the first signs of true Emotional Selling Proposition (ESP) brands. Similar products were perceived as different primarily because of an emotional attachment. Think of Coke and Pepsi. The consumer tends to drink the "label" rather than the cola. During the 1980s the Organizational Selling Proposition (OSP) emerged. The organization or corporation behind the brand in fact became the brand. It was the organization's philosophy that distinguished it from others. For many years Nike subscribed to this form of branding. The internal spirit of the company was so strong that its employees became the main ambassadors for its brand.

By the 1990s brands had gained enormous strength in their own right, and the Brand Selling Proposition (BSP) took over. The brand was stronger than the physical dimensions of the product. Think Harry Potter, Pokémon, Disney, or M&M's. The brand name is found on sheets and toothbrushes, wallpaper and makeup sets. Books and movies aside, the consumer has become more fixated on the brand than the stories.

The world of communication constantly changes. Interaction has become one of the main catalysts. The concept of interactivity has forced us to rethink each and every communication, evaluating and designing it for the ever-demanding consumer. Technological innovation paved the way for MSP brands, which saw consumers taking ownership of their brands. The Canadian brand Jones Soda is a good example of this phenomenon. Consumers design their own label, which Jones Soda guarantees to distribute in the designer's local area. Nike and Levi's websites offer to customize any of their models exactly to your need and size.

## *The Future World of Holistic Branding*

There's every indication that branding will move beyond the MSP, into an even more sophisticated realm—reflecting a brave new world where the consumer desperately needs something to believe in—and where brands very well might provide the answer. I call this realm the HSP—the Holistic Selling Proposition. HSP brands are those that not only anchor themselves in tradition but also adopt religious characteristics at the same time they leverage the concept of sensory branding as a holistic way of spreading the news. Each holistic brand has its own identity, one that is expressed in its every message, shape, symbol, ritual, and tradition—just as sports teams and religion do today.

### An Emerging Global Phenomenon

Will's was a good story, a story that reflected some kind of truth that was out there. But stories need to be substantiated in order to reformulate the way we think. So what started out as an idea became formalized into a research project, and what emerged was a solid methodology that could shape the next generation of brand building.

The very foundation of this book and the theory behind it are a direct result of an extensive research project which sought to determine to what extent the religious factor—faith, belief, and community—could serve as the model for the future of branding. The project investigated the role each of our five senses would play in creating the ultimate bond between the consumer and the brand.

With the advent of globalization, every brand can be accessed from any place on earth with a telephone line. In the absence of a line, the cell phone will pick up the slack. Not a single brand is launched these days without an accompanying Web address.

This is the reality that we took on board in deciding that the project would have relevance only if conducted globally. Our multicultural research team involved people drawn from twenty-four countries, speaking eighteen languages. Additionally, the global research had another objective. We wanted to identify trends, and look at the evolution of local brands to help us create a solid foundation for the implementation of the HSP theory, in order for it to adapt to any market regardless of cultural differences and preferences.

I decided to team up with the global research institute Millward Brown. Their extensive brand knowledge made them an obvious partner for a project of this caliber. *BRAND sense* is the culmination of our extensive study. The idea took seed in 1999 and eventually developed into a global research project that involved some six hundred researchers across most of the globe. It represents the first truly professional brand research effort on a global level.

Research on sensory perception and religious comparisons to branding had never been done before, and we tried our best to remain sensitive to their differences in character, profundity, and ultimate truth. Project *BRAND sense* is therefore a pioneering study. We conducted focus groups in thirteen countries, and performed quantitative tests in three markets. Each country was carefully selected on the basis of market size, brand representation, general product innovations, religious representation, state of brand maturity, and the country's sensory history. We quickly learned that even though a brand is supposedly global, the way local cultures perceive it could be extremely varied. We discovered that the way people use their senses also varies from market to market both in terms of the sensory priority and in sensory sensitivity.

The *BRAND sense* study therefore is a composite of distinct and different markets. For example, we selected Japan, India, and Thailand because all three countries have a well-known history of integrating five senses in their culture and traditions. Some of Japan's most innovative brands often make use of the five senses—for example, when using aroma. The rich design heritage of the Scandinavian countries has made visual identity essential in their communication. The United States and Britain, with their large market size and diverse media, present the biggest challenge in building and maintaining brands. We added countries like Chile, Mexico, Poland, and Spain to our study because of their strong religious and devotional traditions, or because of their history with music and food.

I hope you find *BRAND sense* inspiring and I hope also to introduce you to a new vision for building brands. It's been my intention to draw parallels between existing cases that will help you rethink the way we will perceive brands in the future. But more than that, I hope you will leave the *BRAND sense* experience with a clear idea of how to proceed with creating a multisensory platform for your own brand.

---

### Highlights

The brand building of the future will move from a two-sensory approach to a multisensory approach. The historical development of brands started with the Unique Selling Proposition (USP), where no two products are alike. Succeeding stages included:

- Emotional Selling Proposition (ESP), where products were perceived as different primarily because of an emotional attachment
- Organizational Selling Proposition (OSP), where the organization or corporation behind the brand in fact became the brand
- Brand Selling Proposition (BSP), where the brand was stronger than the physical dimensions of the product
- Me Selling Proposition (MSP), which saw consumers taking ownership of their brands.

The future of branding will embrace the Holistic Selling Proposition (HSP). HSP brands are those which not only anchor themselves in tradition but also adopt characteristics of religious sensory experience to leverage the concept of sensory branding as a holistic way of spreading the news.

CHAPTER 2

# Some Companies Are Doing It Right

**THERE CAN BE NO DOUBT THAT** the effects of branding campaigns are on a steady decline as the cost of reaching consumers in an ever-busier world is on the rise. Household television viewing is increasingly the domain of children. In the United States, the average child is exposed to more than 30,000 television commercials a year, while adults see 86,500.[1] Today's sixty-five-year-old American has on an average watched more than 2 million commercials throughout life—equal to watching TV commercials every day eight hours a day, seven days a week without any break for close to six years.[2]

Given these television-viewing facts, it is not surprising that $244 billion was spent on all advertising in 2003. It is a frenzied world indeed. Every year several thousand new brands appear on store shelves and need to be introduced to consumers. Marketers are facing ever new challenges to whip up the necessary attention required to build these brands. By 2007, 20 percent of U.S. households will have access to systems like TiVo that enable viewers to skip television ads altogether.[3] According to Nielsen Research, the number of men between ages eighteen and thirty-four watching prime-time television has declined by 5 percent. Furthermore, in 2003, 69 percent of U.S. magazines experienced falling circulation figures.[4]

Yet despite the decline in effectiveness, advertising is here to stay. Perhaps the way we communicate our messages needs to be reevaluated to be more closely aligned with today's world. Something new is required to break the two-dimensional advertising impasse. Superb picture quality won't do it. Nor will snazzier graphics. Increasingly, creative ideas set to digital sound aren't the answer either. Incremental improvements have been made in some areas, such as creating ads that mirror the theme of a program, and these to some degree help to keep viewers watching. But no matter what we do, advertising remains a mere flash in the daily life of the consumer.

Suppose we broadened our horizons to encompass as many senses as possible in our messages. Would it work? Skeptics correctly point out that transmitting smell through television is simply a physical impossibility. I would suggest that even though a brand cannot impart an aroma via a television set, there's nothing stopping an aroma from being fully integrated within the brand.

## *Sniffing an Opportunity . . . ?*

You sniff the milk you take from the refrigerator before drinking it; you sniff at the slightest indication of smoke, and then you act accordingly. Our sense of smell keeps us safe by helping us choose fresh and avoid rotten food. Each piece of fruit and cut of meat that finds its way into our shopping cart has passed the sniff-and-feel test, even through its plastic wrap. Instinctively we check for suspicious tears in the packaging and we wait subconsciously for the clicking of the seal when we open a soda or a can of peanuts.

Our senses are more attuned to danger detection than expectations of sensory delight. However, over the past century the advertising world has indulged and catered to our sense of sight in ensuring optimal visual satisfaction. We've become visually sophisticated, and we know that what we see is not always what we get. Be that as it may, the packaging of a product still carries the major load in attracting attention.

If there is a sound, touch, taste, or smell component, well, you'd probably be right in assuming that this is merely a pleasant coincidence. One may ask why these four senses have been neglected and left to their natural protective roles. I cannot think of a single reason!

Almost our entire understanding of the world is experienced through our senses. Our senses are our link to memory and can tap right into emotion. A bright fresh spring day has a particular smell to it. Manufacturers try to bottle this feeling of life's renewal. Then the marketers use the emotional connection to spring to sell their dishwashing liquids, toilet cleaners, shampoos, soaps, window cleaners and, well, you name it.

Bringing on the five senses has worked very well in emotionally connecting people to the rituals of faith. Candles flicker, the incense wafts, the choir strikes up rousing anthems of devotion, there's pageantry, elaborate costumes, and foods for special occasions. Even the sixth sense—the intuitive perception beyond the five senses—is given a special place in the pantheon of world religions.

We store our values, feelings, and emotions in memory banks. Compare that memory to the standard video recorder which records on two separate tracks—one for image, one for sound. The human being has at least five tracks—image, sound, smell, taste, and touch. These five tracks contain more data than one can imagine because they have direct bearing on our emotions and all that they entail. They can fast forward or backtrack at will, and stop just exactly on the right spot in a split second.

Just recently I was walking down a Tokyo street and brushed past someone wearing a distinct perfume. And whoosh! A Pandora's box of memories and emotions immediately spilled open. That whiff took me back fifteen years to my high school days, when a friend wore the exact same perfume. For a brief moment Tokyo didn't exist, and I was back in Denmark, flooded by a warmth of the familiar, and the happy, sad, and fearful times of a high-school boy.

Our memory library begins accumulating material from birth. But this is fluid and flexible, constantly open to redefinition and reinterpretation. When the famous Russian physiologist Ivan Pavlov introduced his famous experiment in 1899, he showed how a dog learns to anticipate food by the sound of a bell. This reflexive behavior extends to humans.

Take the simple example of a bedside alarm. As it signals its wake-up call every morning, you may come to anticipate its sound with foreboding. If, by chance, you hear the exact sound in the middle of the day, it would not be surprising to experience that same sense of foreboding suddenly upon you.

Events, moods, feelings, and even products in our lives are continuously imprinted on our five-track sensory recorder from the second we wake to the moment we sleep. This despite the fact that most mass communication—including advertising messages—that we're exposed on a daily basis comes to us on two of the five available tracks. They're visual and they have sound. We are so used to it, we never give it much thought. Herein lies the anomaly. As human beings, we're at our most effective and receptive when operating on all five tracks, yet not many advertising campaigns, communication plans, or brand-building exercises utilize much more than sight and sound to put their message across.

Do you remember when you bought your first new car? It had a definite new-car smell. Many people cite the new-car smell as being one of the most gratifying aspects of purchasing a new car. The smell is as much a statement of newness as the shiny body.

In fact there is no such thing as a new-car smell. It's an artificial construct, a successful marketing ploy that taps directly into fantasy. This smell can be found in aerosol canisters on the factory floor that contain that "new-car" aroma. As the car leaves the production line, the scent is sprayed throughout its interior. All in all it lasts about six weeks, and then is overtaken by the rough and tumble of dirty track shoes, old magazines, and the empty coffee cup you drank from on your way to work.

Ironically, neither the odometer nor your efforts at extreme tidiness can define when your car is no longer "new"—it's the disappearance of that new-car smell that creates the demarcation between a new and an ordinary everyday item. Of course you can prolong the sense of newness by stepping into your local car accessory store and buying a can of the new-car smell yourself!

We are surprisingly unaware of the way our senses interact with our day-to-day experience. Bondi Beach in Sydney is fringed with stores all selling the usual summertime equipment. Umbrellas, sarongs, boogie boards, sun creams, and sodas. On a cold winter's day, when a rough southerly was blowing, a friend of mine needing to buy a last-minute birthday present popped into one of these stores to peruse the jewelry. She suddenly found herself browsing the rack of swimsuits. Surprised by her own behavior, she slowly became aware that the air seemed filled with summer even though the swimming season was a good five months away. She jokingly asked the sales staff to reveal their unseasonable secrets. Taking her into their confidence, they led her to the corner of the store, where a concealed machine consistently pumped out the subtle smell of coconuts. She didn't buy the swimsuit, but a week later she booked a trip to Fiji!

The power of suggestion can be found everywhere. Rice Krispies that don't snap, crackle, and pop are quite simply considered to be stale, even though the taste has not changed and they may still be perfectly good to eat. As for cornflakes, Kellogg's considers the crunchiness of the grain as having *everything* to do with the success of the breakfast product. Emphasis is placed on the crunch we hear and feel in our mouths rather than the sound effects we hear on commercials.

Kellogg's has spent years experimenting with the synergy between crunch and taste. As part of this research they made contact with a Danish commercial music laboratory that specializes in the exact crunchy sensation of a breakfast cereal. Kellogg's wanted to patent their own crunch, and trademark and own it in the same way they own their recipe and logo. So the laboratory created a highly distinctive crunch uniquely designed for Kellogg's, with only one very important difference from traditional music in commercials. The particular sound and feel of the crunch was identifiably Kellogg's, and anyone who happened to help himself to some cornflakes from a glass bowl at a breakfast buffet would be able to be recognize those anonymous cornflakes as Kellogg's.

The day Kellogg's introduced their unique crunch to the market, their brand moved up the ladder. They'd expanded the perception of their brand to incorporate four senses (including touch) rather than the more limited sight and taste. So by appealing to another of our five senses they broadened their brand platform.

Expanding your brand platform to appeal to as many senses as possible makes sense. Think of walking past a bakery where you can smell the aroma of the warm bread, without stopping. It's extremely difficult. In the supermarkets of Northern Europe freshly baked bread is prominently displayed near the entry to the store. Although there's no immediate evidence of a bakery, if you look carefully at the ceiling, you will spot vents that are specially designed to disperse baking aromas. It has proved a profitable exercise in increasing sales—not only of baked goods, but across many product lines.

What aroma do you most associate with cinemas? I doubt that it's the smell of celluloid or other people. Chances are you're thinking of popcorn. In fact, the smell of popping corn has become so strongly linked with going to the movies that if it wasn't there you would more than likely feel an unidentifiable absence.

The unique aroma of popcorn, the texture and sound of crunching cornflakes, or the distinctive smell of a new car has very little to do with the actual product, or for that matter its performance. Yet these components have come to play an almost fundamental role in our relationship with these products. These forms of sensory stimulation not only make us behave in irrational ways, but also help us distinguish one product from the next. They've embedded themselves in our long-term memory and have become part of our decision-making processes.

It is these very processes that point the way toward the next generation of brand building. Over the next decade we will witness seismic shifts in the way we perceive brands. It can be compared to moving from black and white or color television with mono sound to high-definition color screens installed with surround sound.

Look at the page you're now reading. All you see are black letters printed on a white page. That's all I have at my disposal to convince you of a world that can be enhanced not just by vision, but by every other sense as well. Imagine a world devoid of color where everything we see is in black and white. Then try to explain the color red to a person who has only black-and-white vision. It's an enormous challenge, and one no different from the challenge facing brands, because ultimately they will have to move from the safety of their two-dimensional track and contemplate how to navigate in a world of color. It's a giant step for the advertising world, but an essential leap if they are going to be players in this new arena of sensory experience

The game has already begun. In fact, as far back as 1973 Singapore Airlines broke through the barriers of traditional branding with their Singapore Girl, a move that would prove so successful that in 1994 the Singapore Girl celebrated her twenty-first birthday and became the first brand figure to be displayed at the famous Madame Tussaud's Museum in London. Previously airlines had based their promotions on cabin design, food, comfort, and pricing—ignoring the total sensory experience they could offer. Singapore Airlines made the shift when they introduced a campaign based exclusively on the emotional experience of air travel.

With a brand platform emphasizing smoothness and relaxation, their strategy was to move away from portraying themselves simply as an airline and instead to present themselves as an entertainment company. In the process, a new set of brand tools were invented and introduced. The staff uniforms were made from the finest silk in a fabric design based on the patterns in the cabin décor. The staff was styled right down to their makeup. Flight attendants were offered only two choices of color combination based on a palette designed to blend with Singapore Airlines' brand color scheme.

The Singapore Girl was turned into an icon. The selection criteria for staff are inflexibly stringent. Cabin crew members have to be under twenty-six years of age, and the first hurdle a woman faces is fitting her body perfectly into the one-size uniform. Her beauty has to compare to the models used in the enticing ads. Above and beyond the usual training for flight attendants, she not only has to look the brand, but she has to act the brand as well. This includes strict instructions on how to speak to passengers, how to move in the cabin, and how to serve food.

Politically incorrect as this may appear in some countries, Singapore Airlines was clearly driven by an aim to establish a true sensory brand experience encompassing so much more than what the passengers could see and hear. Even the announcements from the captain were carefully scripted by the advertising agency.

The sensory branding of the Singapore Girl reached its zenith by the end of the 1990s, when Singapore Airlines introduced Stefan Floridian Waters. Not your average household name, to be sure, Stefan Floridian Waters is an aroma that has been specifically designed as part of the Singapore Airlines experience. Stefan Floridian Waters formed the scent in the flight attendants' perfume, was blended into the hot towels served before takeoff, and generally permeated the entire fleet of Singapore Airlines planes. The patented aroma has since become a unique trademark of Singapore Airlines.

Of course it is often difficult to identify a particular odor, and even more difficult to describe it in words. Those who recall the unique smell of the airline cabin describe it as smooth, exotically Asian, and distinctly feminine. If you were to ask travelers who take a subsequent journey with Singapore Airlines about this unique scent, they all report instant recognition upon stepping into the aircraft. It's a smell that has the potential to kick-start a kaleidoscope of smooth comfortable memories—all reflecting the Singapore Airlines brand.

## *Brand Bland*

It was less than fifty years ago when the first documented evidence on the positive effects of branding appeared. During the late 1950s it appeared that consumers were prepared to pay more for branded products—even if the nonbranded item was of the same quality, appearance, and taste. Most, if not all, of the knowledge we have today about branding has its roots in the 1950s and 1960s.

The intense focus on building a brand around its "personality," namely, giving the brand values and feelings in order to distinguish it from the next, evolved in the 1970s and 1980s. There have been no earth-shattering changes in our perception of brands in all this time. Even the World Wide Web, an interactive medium, still primarily contains banner ads as its primary advertising tool—despite the fact that those ads lack a true interactive foundation.

There is no doubt that the marketing community is technologically and creatively smarter in the execution of television commercials, print ads, billboards, and radio promotions than in past years. But as we have previously observed, all the communication techniques used today have one thing in common: they're all based on two senses—sight and sound. This flies in the face of the fact that human beings have three more senses that can be addressed. Furthermore, research shows just what a large role our olfactory capabilities play in our decisions.

## *Do You Hear Me, Hear Me, Hear Me . . .*

Repetition has been one of the most prominent techniques used by advertisers to ensure a message is understood and remembered by the consumer. A classic television campaign will be seen or heard by the consumer, on average, three times. This applies across the board, no matter where the consumer has access to commercial television. In a fact that barely needs remarking upon, the more often a message is repeated, the better it is remembered. As is the brand.

There clearly is a limit to just how many times we can repeat what we do. To what extent can we saturate the airwaves with messages and still expect people to pay attention? Watch any news station and the screen is packed with news bars, stock ratings, breaking-news updates, as well as the TV presenter—and all this happens at the same time on one screen.

The reality is that people are spending less time in front of television, less time reading magazines, and less time listening to the radio. Even so, over the past five years, advertising spending has increased on average by 8 percent annually.[5] By the same token, the average consumer is exposed to 9 percent more commercial messages each year.[6] In 1965 the average consumer remembered 34 percent of the ads shown on TV. In 1990 he remembered only 8 percent.[7] More and more money is being spent executing increasingly less effective brand campaigns. In short, advertising has hit a brick wall.

## *Replacing Repetition with Sensory Synergy*

Let's look at movies as an example. Remove the dialogue, the sound effects, and the music and I'm sure you'd agree that we're not left with much that entertains us. Conversely, remove the images and listen to the soundtrack. Again, hardly stuff that will keep us enthralled. The value of movie entertainment is the combination of audio and video working together. Only then do we have cinematic magic. Magic enough to make a 2 + 2 = 5 equation work!

To some extent we have reached a level where we can say that the positive synergy is effective just by forming this equation. But at what point is this enough? Would it be true to say that if we were able to combine this workable synergy with taste, touch, and smell we'd add another hugely substantial dimension? Could the formula be as simple as: sound + vision + touch + smell + taste equals 2 + 2 + 2 + 2 + 2 = 20? Would we then discover a positive synergy across and between each of our five senses? Is it the case that delicious-smelling food tastes better? Or a heavier mobile phone implies a better quality? Perhaps perfume smells better if it is presented in a stylish bottle? Will a brand have more value if it imparts a sense of smell, touch, and taste in addition to the audio and visual aspect?

There may be something to it. New York–based *Condé Nast Traveler* magazine has named Singapore Airlines the best airline. And almost every other independent study has concurred: Singapore Airlines leads the way. This despite the fact that their food is average and their leg room is no better than many of the other airlines that rank in the top twenty.

To what extent can we regard the increased sales in a supermarket pumping the smell of freshly baked bread through their aisles as coincidence? How can we explain why the crispiness of our breakfast cereal makes it taste fresher and better?

By the end of the 1990s Daimler Chrysler established an entirely new department within the company. This was not to design, build, or even market cars. Its job was solely to work on the sound of their car doors. That's it. With a team of ten engineers allocated to the task, their only role was to analyze and then create the perfect sound of an opening and closing door.

Over the years, car manufacturers have learned a lot about what sells cars. And it's not what we think it is. It's not necessarily the car's design or even the acceleration. Studies show that the interior design—including the way the doors open and close—helps determine the choice. The interior design is extremely important because generally women relate more to the feel of the interior than they do to the features outside. So the way the doors close can be an important factor in the perception of quality. Daimler Chrysler understands this.

The days when people could discuss the true value of brand building are long gone. Models have been established, proven, and enhanced a million times over. We have now reached a point where another dimension to building brands needs to be added. This will help ensure that branding won't be killed by its own success.

## *The Five-Dimensional (5-D) Brand*

Try this exercise. Draw a sensory model (a sensogram) of your brand, and then describe its appeal in both visual and audio terms. Although this will by its very nature be a fairly subjective assessment, I suggest that you approach it using the following criteria:

- How clear is the image?
- How distinct is the image?
- Do consumers perceive the image of your brand consistently?
- Is the image a memorable one?

### Sight

Sight is the most seductive sense of all. It often overrules the other senses, and has the power to persuade us against all logic. Consider the food and color test that Dr. H. A. Roth performed in 1988. He colored a lemon-and-lime flavored drink in various degrees of intensity. He then asked hundreds of students to say which was the sweeter. Most got it wrong. They believed that the stronger the color, the sweeter the drink. But in fact it was quite the opposite: the stronger the color, the more sour it actually was![8]

In another test, C. N. DuBose asked the subjects to taste grape, lemon-lime, cherry, and orange drinks. There was no trouble correctly identifying the flavor if the color matched. But when color and flavor were switched, only 30 percent of those who tasted the cherry could identify the flavor. In fact, 40 percent thought the cherry drink was lemon-lime.[9]

Vision is all about light. As early as the fifth century BC the Greeks recognized the link between the eye and the objects seen. By the fourth century BC Aristotle rejected the idea that a visual fire emanated from the eye, reasoning that if vision were produced by fire in the eye, we would see in the dark. The difference between our day and night vision is that our night vision is color-blind.

One of the most important art movements in history occurred when a collection of artists in nineteenth-century France began to study the effects of light. They became known as the Impressionists, and their work is essentially a study of the effects of changing light on any given object. They took their oils and their easels outdoors and painted haystacks and water lilies and the like, over and over again, recording different times of day as well as different seasons. To truly see what an artist saw when painting a given picture, you should look at it under the same light.

Within the range of visible light, different wavelengths appear to us as different colors—therefore most colors that we see are composed of a range of wavelengths. It is not surprising that the eye has been understood to work like a camera—its function is to send a perfect image to the brain. This misperception is so widespread that it even has a name. It's called the homunculus fallacy (homunculus is Greek for "little man"). The fallacy is the idea that when we see something, a small representation of it is transmitted to the brain to be looked at by a little man.

The function of the visual system is to process light patterns into information useful to the organism. We have surprisingly low visual acuity (resolution) in parts of the visual field that are not at the center of where we are looking—the center of gaze. We are not aware of this, because we usually move our center of gaze to whatever we want to look at.

Light passes through the pupil, and the lens focuses the image on the retina, a sheet of layers of neural tissue that lines the back of the eyeball.

There are photoreceptors in the first layers of the retina which have light-absorbing chemicals. The signals pass through the first layer to the ganglion cells, which send their signal from the eye via the optic nerve to the brain. This then translates into what we see.

And we all see differently. Half-full or half-empty. You say orange, I say vermilion. Sight, one might say, is truly in the eye of the beholder—which is why companies like the renowned authority Pantone specialize in developing tools to help designers communicate colors.

**THE VISUAL BRAND** Let's examine the Coca-Cola brand visually. It has a very clear sense of color. Quite simply, wherever there's Coke, there's red and white. Coca-Cola takes its colors extremely seriously. Santa Claus traditionally wore green until Coca-Cola began to promote him heavily in the 1950s. Now in every shopping mall across the western world, Santa wears the colors of Coke. The consistent use of the colors, the dynamic ribbon, the typography, and the logo have established a very clear and unambiguous image which has survived for decades and is memorable to anyone who has been exposed to the brand. It's a brand that will, without a moment's hesitation, earn full marks for its visuals.

## Sound

I have a memory from when I was a child. We were asked to sit silently in a circle on the floor and do nothing but listen for two minutes. There were no eventful sounds, mostly silence. Yet when asked, we all could recall hearing something. And each of us heard something different. For some it was a cough, for others it was a footstep, or a door slamming. Traffic. Leaves rustling. I heard a clock ticking.

Children have more acute hearing than adults. They can recognize a wider variety of noises—and memorize these more easily. As we grow older we lose our sensitivity, unless of course we constantly exercise our listening faculties. But even the most sensitive human ear is not as finely tuned as a dolphin's. Dolphins' hearing is fourteen times better than humans.

As smell is connected to memory, so sound is connected to mood. Sound does in fact generate mood. It creates feelings and emotions. A love movie isn't nearly as emotional if you watch with the sound off. Sound can inspire joy and sadness in equal measure.

Like our other sense organs, our ears are extremely well designed. They serve two very important purposes. Besides hearing sounds, our ears maintain our balance.

Sound originates from the motion or vibration of an object—just like the vibrations of a drum. This motion sends vibrations or sound waves through the air, in the same way that ripples form on a pond. The outer ear funnels these vibrations into the ear canal, where they move by a process similar to Morse code until they hit the eardrum. This sets off a chain of vibrations. The eardrum vibrates against the three smallest bones in the body, moving the sound through an oval window into the labyrinth, a maze of winding passages. At the front of the labyrinth is a coiled tube resembling a snail's shell. Here the 25,000 receptors pick up the signals and send them to the brain, and so we hear. Balance is controlled at the end of the labyrinth.

It appears that loss of hearing is worse than loss than the loss of sight. For example, in a letter she wrote in 1910 Helen Keller said, "The problems of deafness are deeper and more complex, if not more important, than those of blindness. Deafness is a much worse misfortune. For it means the loss of the most vital stimulus—the sound of the voice that brings language, sets thoughts astir and keeps us in the intellectual company of man."[10]

**BRAND SOUND** The second dimension that is heavily leveraged in today's brand-building process is the use of audio. Despite the fact that audio technology has been available for over a hundred years, the use of audio has not been perfected nearly to the degree of its visual counterpart.

Using the same criteria we applied to a brand's visuals, Intel stands out as the company with the clearest, most distinct, consis-

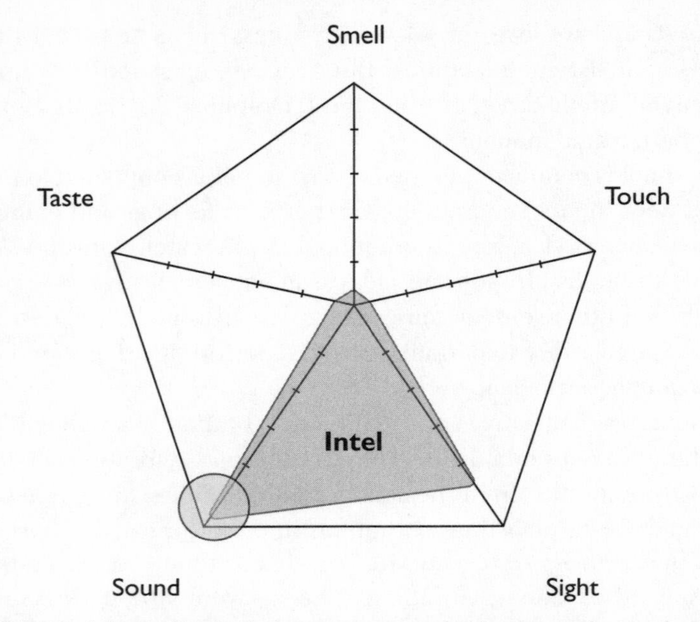


**FIGURE 2.1** A brand like Intel manages to score highly on sound even though the brand's core offering has nothing to do with audio. The criterion for achieving a perfect score is to have a distinct and memorable sound.

tent, and memorable use of sound. The Intel Inside tune has been around since 1998, making the invisible (the chip) visible via the short, distinct sound used throughout all of Intel's advertising and brand-building campaigns. Research shows that the Intel jingle, also known as the wave, is as distinct and memorable as the Intel logo. In fact, studies have shown that in many cases people remember the Intel wave better than the logo.

## *That's It . . . At Least According to Conventional Wisdom*

Up until now, that would be it for building a brand. Perfect visuals. Perfect audio. This is where the brand-building process traditionally stops. It stops despite the fact that a total sensory experience would at least double, if not triple, the consumer's ability to memorize the brand.

If you decided to examine any of today's Fortune 500 brands, you would quickly realize that decades concentrating on the audiovisual dimensions has narrowed the focus to only these two dimensions, neglecting the other senses as if they do not exist. I'd even go so far as to say that the perfect utilization of audio has not yet been achieved.

So many brands focus their energies on strong visuals, often to the detriment of the audio component. Visit the Internet sites of the world's Fortune 500 brands, and you will quickly come to realize that only 4 percent use sound as an integrated element online. Furthermore, only 9 percent of all brands utilize the strengths of audio in making their brand more distinct, clearer, consistent, and memorable across a majority of their channels.

Consider this. When you open a bottle or can of soda, there's a distinct sound. No one has thought to brand this. Then there's Microsoft's start-up notes. Yet Microsoft changes them each time they release a new version of their operating system. One wonders why Porsche hasn't branded a new-Porsche smell. Why doesn't Motorola have a Motorola ringing tone? After all, 15 percent of the world's mobile phone users listen to their Motorola phone ringing approximately nine times a day.

Jean-Martin Folz, the chairman of the French car manufacturing giant PSA Peugeot Citroën, decided to adopt a sensory branding strategy to create a group identity around the two very distinct brands. At the same time that the company created a corporate identity, the group developed a sound identity. Interestingly, this sound identity was used internally, rather than for commercial purposes. Each morning, when 65,000 employees turn on their computers, they are greeted by the group's signature tune rather than Microsoft's start-up notes. The PSA group's tune was also applied to their telephone "hold" music. It went further. When Folz gave a speech about the group's strategy at the 2003 Paris auto show, the music was played as an introduction, before he took to the stage.

## Smell

You can close your eyes, cover your ears, refrain from touch, and reject taste, but smell is part of the air we breathe. It's the one sense you can't turn off. We smell with every breath we take, and that's around 20,000 times a day. It is also the sense we most take for granted. There's no cultural activity that caters to it—no sniffing galleries, no concertos written to surround us with odor, no special menu of smells created for grand occasions, and yet . . . it is the most direct and basic sense.

Observe an animal in a new place. The very first thing it does is sniff around. Odors give them most of the information they need to gauge their potential safety. Smell is also extraordinarily powerful in evoking memory. Where you may be at a loss to conjure up the details of your childhood home, a whiff of homemade bread can instantly transport you back in time. As Diane Ackerman says in her poetic study *A Natural History of the Senses*, "Hit a tripwire of smell, and memories explode all at once. A complex vision leaps out of the undergrowth."[11]

No one has managed to describe the nose more elegantly than Lyall Watson. In *Jacobson's Organ,* his comprehensive and idiosyncratic study of smell, he refers to smell as a "chemical sense." He goes on to explain, "Receptor cells in the nose translate chemical information into electrical signals. These travel along olfactory nerves into the cranial cavity, where they gather in the olfactory bulbs. These, in turn, feed the cerebral cortex, where association takes place and nameless signals become transformed into the fragrance of a favorite rose or the musky warning of an irritable skunk."[12]

Smell is almost impossible to describe. We are exposed to thousands of different smells yet we have an extremely limited vocabulary to address them. Watson points out how scant the vocabulary for auxiliary odors (such as the way a home or a cupboard might smell) is in every culture. "In Central Africa alone, auxiliary odors are described as phosphoric, cheesy, nutty, garlicky, rancid, ammoniac and musky."[13] We often "borrow" from the wider vocabulary of food and taste to describe a scent.

How we perceive body odor is culturally determined. Some Mexicans still believe that the smell of a man's breath is more responsible for conception than his semen is. In Japan, 90 percent of the population has no detectable underarm odor, and young men who are unfortunate enough to belong to the smelly minority can be disqualified from military service on that ground alone. Napoleon had no such problem. He wrote to Josephine, "I will be arriving in Paris tomorrow evening. Don't wash."[14] George Orwell did not share Napoleon's passion, and writing almost a century later, stated that "The real secret of class distinction in the West can be summed up in four frightful words . . . *the lower classes smell.*"[15]

Smell has played its part in war. Jack Holly, a U.S. Marine who led patrols in Vietnam, says, "I am alive because of my nose. You couldn't see a camo bunker if it was right in front of you. But you can't camouflage smell. I could smell the North Vietnamese before hearing or seeing them. Their smell was not like ours, not Filipino, not South Vietnamese, either. If I smelled that again, I would know it."[16]

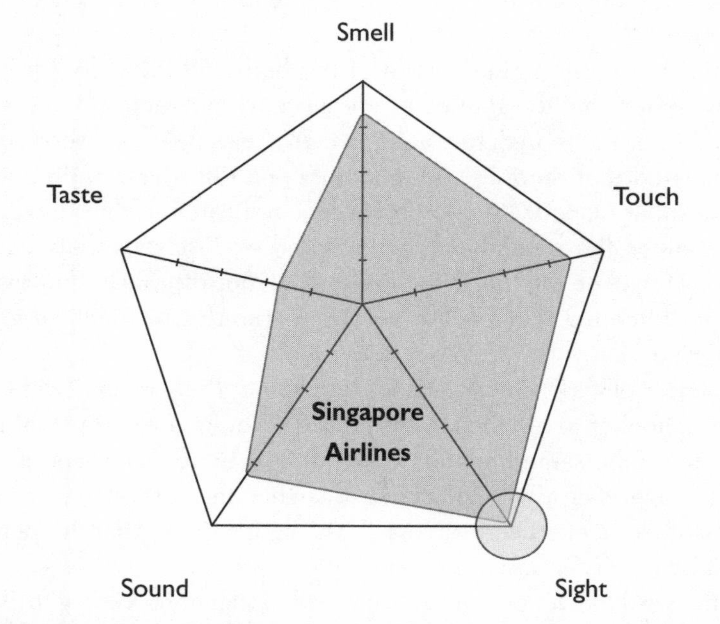


**FIGURE 2.2** Singapore Airlines scores high on the smell dimension in comparison with any othr airline brand.

**THE BRAND SMELL** If you agree with me this far, I'm sure you'll be even more surprised to learn how few brands have established a distinct aroma; in fact, less than 3 percent in the Fortune 1000 list have even given it a thought.

But just as your brand's sight and sound need to be clear and distinct, so does its smell. All it needs to be is a subtle scent that in some cases is so fully integrated with the brand that you'd hardly notice it.

## Touch

Touch is the tool of connection for those who have the misfortune to be both blind and deaf. When all else fails, the skin can come to the rescue. Such was the experience of Helen Keller, who became deaf and blind through illness at age two. The unruly child was dragged to the water pump by her teacher, who held her hand under the stream while signing W-A-T-E-R into her palm. This marked the beginning of an arduous but rewarding journey that ultimately led to literacy and opened up a world of Braille and books that could be read by touch.

The skin is the largest organ of the body. Additionally the elements comprising the skin have a large representation in the cortex of the brain. We're instantly alert to cold, heat, pain, or pressure. It is estimated that there are 50 receptors per 100 square millimeters each containing 640,000 microreceptors dedicated to the senses. As we get older, these numbers decrease and we lose sensitivity in our hands. However, our need for touch does not diminish, and exists beyond detecting danger. We need the stimulus of touch to grow and thrive.

A series of experiments was undertaken at the University of Colorado School of Medicine by Dr. John Benjamin. Two groups of rats were given the same tools for survival—food and water, and a safe living space. The only difference was that the rats in one of the groups were stroked and caressed. The results were that the petted rats "learned faster and grew faster."[17]

The word *touch* encompasses a world of meaning. We try to "stay in touch" with friends, and we "lose touch" with some. People are partial to the "personal touch" as an expression of a personal idiom.

We feel "touched" by gestures of care and concern, and express distaste by refusing to "touch it with a ten-foot pole." We're "touched" by madness or a bit of the sun. The list goes on.

Touch alerts us to our general well-being. Pain travels from skin to brain and triggers warning systems that demand attention. Those who feel no pain may sustain serious injury without being aware of danger. A therapeutic touch can also ease pain. Massage has long been a prescribed remedy for tense muscles and poor circulation in Asian countries and has more recently expanded to the West. Preachers lay hands on those who need to be healed. The Japanese have mastered shiatsu, a type of acupuncture using the fingers.

From the parent's touch of a child to the sensual caress of lovers, touch is ultimately the true language of love.

**THE BRAND TOUCH** What's the texture of your brand? For many companies this would not apply, but still close to 82 percent of all

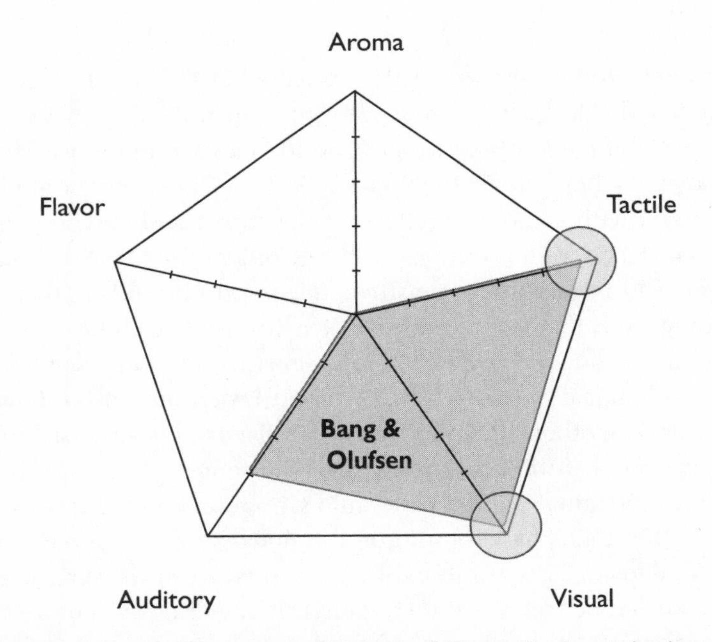


**FIGURE 2.3** Bang & Olufsen scores high on touch. The tactile element has become one of the core brand elements differentiating B&O from its competitors.

brands appearing on the Fortune 1000 list would be able to leverage this if they were made aware of it.

One of the most distinct brands that appeal to the tactile sense is the luxury electronics company Bang & Olufsen. Since their products first appeared in 1943, Bang & Olufsen have put as much detail into their design as they have into the quality of their sound. One of their many innovations has been the all-in-one remote control—enabling the user to use the same device for the television, the radio, the CD, the tape recorder, as well as the lighting in each room. This invention, which first appeared in 1985, has evolved to become a streamlined sensual piece of equipment that oozes quality. Other companies have introduced similar pieces of equipment, but the Bang & Olufsen remote is heavy, solid, and quite distinct. This sense of gravitas is duplicated across every Bang & Olufsen product line, from telephones to speakers, including earphones and the whole range of accessories.

## Taste

Taste is detected by special structures called taste buds. It is generally believed that girls are more sensitive to taste than boys. The belief is well founded because girls do in fact have more taste buds than boys. We have about 10,000 taste buds, mostly concentrated on the tongue, with some at the back of the throat and on the palate. Everyone tastes differently. As you get older, your sense of taste changes, and becomes less sensitive, making it more likely that you will enjoy foods that you considered "too strong" as a child.

There are four types of taste buds, sensitive to sweet, salty, sour, or bitter chemicals respectively. Different taste areas of the tongue are better than others at detecting certain flavors, because each type is concentrated in different regions of the tongue. The very tip is best at sweet things (noted in a child's preference to lick a candy sucker rather than chew it), sour on the sides, bitter at the back, and saltiness all over. Taste is formed from the mixture of these basic elements. Different tastes are distinguished by various combinations and a more sophisticated sense of smell.

I remember a school excursion to a large snack factory in Denmark. As we walked by the shaped corn rings, I snuck a few into my mouth. I expected that familiar cheesy flavor but was astounded when I tasted nothing. Absolutely nothing. All I experienced was a strange texture in my mouth. Then I discovered that the snack had yet to undergo a flavoring process before it was packaged. To this day I remember the awful blandness, and it has served to remind me of the importance of taste.

Those who cannot see are blind. Those who cannot hear, deaf. Those who cannot speak, mute. But those who cannot smell or taste are left hanging: they suffer from an absence without a name. Is "taste-blind" the condition that doesn't warrant a term? Food is an integral part of life. Social interaction happens around the table, and food plays a vital role in tradition and ritual. You would still partake in the intimacy of sharing a meal, but the pleasure would be lessened.

The result of losing the sense of taste often generates deep depression. A friend of mine suffering from this phenomenon told me that she would live without any of her senses—but taste. Taste goes hand in hand with smell, and for those who cannot taste, all that is left of a gourmet meal is the texture and temperature of the food. Those who suffer such a loss say that it is like forgetting how to breathe. We take smell and taste for granted, unaware that everything around us has a smell—until, that is, nothing smells at all.

Taste and smell are closely related. It would not be incorrect to assume that one smells more flavors than they taste. When the nose fails, say from a bad cold, taste suffers an 80 percent loss. Loss of taste without loss of smell is pretty rare. Full sensory appreciation of food also involves its appearance, its consistency, and its temperature. The author of a respected British medical journal says that if doctors got closer to their patients, they could smell the ailment. He believes that certain illnesses produce certain odors: a patient who smells like whole-wheat bread may have typhoid, and an apple smell just may indicate gangrene.[18]

Most of the descriptive terms and phrases we have for smell are associated with food. Smell is estimated to be 10,000 times more sensitive than taste, making taste the weakest of our five senses.

**THE BRAND TASTE** Apart from the food and beverage industry, taste is a tricky sense for most brands to incorporate. However, brands that can incorporate taste can clearly build a very strong brand platform. In fact, close to 16 percent of the Fortune 1000 brands could add taste to their brand platform, yet almost none have so much as given this a cursory glance.

Colgate is one of the exceptions. They've patented their distinct toothpaste taste. It's important to note that they have not to date extended this distinctive taste to their other products, like their toothbrushes or dental flosses. So although they've been totally consistent with establishing the Colgate "look" across their product lines, they've been inconsistent by not building their unique taste into products other than toothpaste.

Despite this lack of consistency, Colgate probably ranks as one of the best brands in applying a distinct taste to its product, although there still remains a fair bit of room to leverage taste as part of the brand's extension strategy.

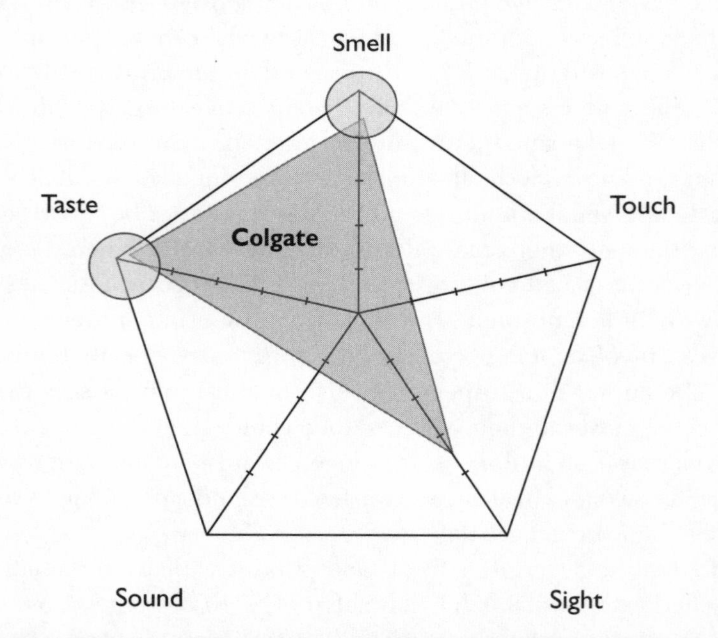


**FIGURE 2.4** Despite its failure to leverage the taste across Colgate brand extensions, the Colgate taste surpasses most other brands that fall outside the food and beverage category.

The taste of Colgate toothpaste, the well-designed Bang & Olufsen remote control, the Intel digital sound wave, and Coca-Cola's distinctive red and white have one thing in common: they've all created a third dimension to their product. Their strong sensory uniqueness is distinct enough for users to recognize without the usual logo or typography cues.

## *Introducing the Signature Brand*

Have you ever heard of a "signature dish"? The term is used by chefs when designing a specific dish which they hope to become known for. In time they may enhance it, add a little spice here, or an herb there, but it's the entrée with which they're associated.

This phenomenon is fascinating not only because it allows chefs to create their own niche in a highly competitive market, but also because it leads to other dishes related to the primary or "signature" dish. People will return to the restaurant because they know that everything on the menu will be in harmony with the signature dish. Everything about the surroundings will also play a role: The décor, the manner in which the food is presented, the plates, the feel of the cutlery, and the attitude of the staff. The reputation of the food is only part of the sensory package. What makes a dish truly memorable is the synergy that exists between the different elements of the whole sensory package. If the chef were appealing only to taste and aroma, it is doubtful if the restaurant would achieve the same results.

The effects of sensory branding are astounding. Yes, it's possible to create a truly spectacular commercial, or an impressive advertising jingle, but they begin to become effective only when the two elements are combined. The effect is magnified many times over when you include any of the other senses.

This total sensory synergy produces a domino effect. In the way impressions are stored in the brain, if you trigger one sense, it will lead to the another, and then another . . . a whole vista of memories and emotions can instantaneously unfold. Succeeding with two elements is only half a story; creating a synergy across the senses should be the ultimate goal.

## *Moving Toward a Multisensory Model*

So the idea of sensory branding sounds good in theory? Well, practical steps need to be taken in order to move your brand from its two-dimensional world into a five-dimensional place. There are strategies to employ so that this transition will be successful. The creation of a sensory brand is a six-step process. Each step is designed so that you won't lose control of the brand. It will ensure that you don't misrepresent the brand, and most importantly you won't end up with a situation where the brand does not fulfill the promises it makes.

### Setting the Stage

To develop a successful sensory branding strategy it is important not to plunge right in and start adjusting the sound, smell, and touch of your brand. Before a chef touches his ingredients, he needs a clear vision of the type of gastronomic journey he wants for his customers. When formulating a sensory branding strategy, one of the key criteria for success is to set the stage, which will allow you to carefully select the channels, the tools, and the senses you intend to tap into when building your future brand.

Setting the stage is an apt metaphor. Each element of your brand contributes to the entire show. The first questions we need to ask are what exactly we wish to create in our brand theater and what message we need to impart. It's essential to be perfectly clear about this core message before we can take the next step.

### Smashing Your Brand

In 1915 a designer from the Root Glass Company of Terra Haute, Indiana was asked to design a glass bottle. The brief was pretty straightforward: The bottle design should be so distinctive that if it were broken, the pieces would still be recognizable as part of the whole. He succeeded beyond his wildest dreams. The bottle he designed was the classic Coke bottle, which has become one of the most famous glass icons ever. The bottle is still in service, still recognizable, and has been passing the smash test for every generation over the last eighty years.

The Coke bottle story reveals a fascinating aspect from a brand-building perspective, because in theory all brands should be able to pass this sort of test. If you removed the logo from your brand, would it still be recognizable? Would the copy stand up to it? Would the colors, graphics, and images standing alone pass the test?

Can your brand survive being smashed? It is an interesting exercise that can remove a logo-fixated mind-set and bring you closer to a philosophy valuing all elements that create your brand. Two black ears from a well-known mouse are instantly recognizable as Disney. A Singapore Girl suggests Singapore Airlines. These are only components of the brand, and yet they're unmistakable. The trick is to create each element so that it's so strong, so able to stand alone, yet at the same time so integrated and synergistic that it can take the brand to a whole new level of familiarity.

## Understanding the Brand Ingredients

In order to successfully smash your brand, you need to have an intimate understanding of what it's actually made up of. What are the drivers behind the visual strategy? What is the theory behind the sound? What role does aroma play in the message? How can you convey its tactile sense on a television screen? How does it taste?

To go back to our chef analogy, we need to know the value of each ingredient. We need to know what works with what, and how to handle and prepare each ingredient in order to achieve the perfect mix and create the optimal synergy for our five senses.

## Gathering the Pieces

Once you've set the stage for your brand, and broken it down to its various parts, it's time to gather the pieces. In putting it together again, you will no doubt be much more familiar with the various parts. This familiarity will allow you to reconstruct your brand in such a way that each sensory component is enhanced and can stand alone.

### Releasing the Brand

So far the number of true sensory branding cases across the globe can be counted on one hand. Part of the *BRAND sense* project has been to explore the details of what makes a successful multisensory brand. Does a five-sense synergy really exist? What combinations of senses work best together? How do you transfer an emotion communicated by one sense to another ensuring a positive synergy? How do you measure the success of this approach?

The questions are endless, and with support from the global research institute Millward Brown, I've attempted to answer them. The *BRAND sense* project has primarily been concerned with finding the most effective ways to transform theory into a practical sensory branding approach.

### The Art of Selling Perception

Building brands requires building perception—nothing more, nothing less. Creating the perfect perception requires the perfect sensory appeal. The aim is to help you revitalize your brand by evaluating and optimizing every dimension that contributes to the perception of your brand. Each aspect of this journey will be examined in detail beginning with the next chapter, where I'll start by smashing your brand. By the end you will certainly be able to establish a totally new platform for your Brand 2.0.

## *Case Study: Navigating Uncharted Brand Channels*

Nokia currently boasts an astounding 40 percent *global* market share in mobile phones. This translates into 400 million people who use their Nokia phone on a daily basis. Besides the more obvious elements that characterize Nokia phones, their less obvious branding tools have created what Nokia has become today. According to the brand consultants Interbrand, Nokia is the world's eighth most valuable brand, and it is estimated to be worth $29 billion.

## Read Chinese with Nokia

The sound language of Nokia is only part of their invincible brand story. Add their navigation and user interface and you realize just how conversant you've become with the Nokia brand.

A couple of months ago a friend of mine set out to fool me by changing the display language on my Nokia phone from English to Mandarin. At first I was taken by surprise when every icon on the screen was written in Chinese characters. However, my familiarity with the Nokia system was such that it was as if I could actually read Chinese. I intuitively found my way to the language function, where I could reprogram it back to English. The language of choice played almost no role in my ability to navigate. It was in fact the Nokia language, which seamlessly carried me across the cultural divide.

This is a situation that only a serious market leader can create. Nokia has established its position through consistent education of the consumer. Nokia users are thoroughly familiar with the interface. Any Nokia user can find the most vital functions on his cell phone without even looking. This, you may say, is more luck than calculated loyalty building. Not so. Think about it. How often have you been frustrated by a new video machine, microwave, or dishwasher? Even if you've purchased a familiar brand, the new operating system often proves so challenging that you are reluctant to even attempt it. The effort proves too demanding.

Habit plays a large part in generating brand loyalty. This is a fact that you may not even be aware of. One of the *BRAND sense* surveys asked respondents to choose between Nokia and Sony Ericsson phones. One respondent clearly admired the Sony Ericsson for its light weight and stylish features, but he chose Nokia because it simply felt easier to use. This despite the fact that the Sony Ericsson was cheaper, had more features, and was more stylish.

## Nokia Knows that Laziness Builds Brands

In contrast with almost every other cell phone manufacturer, Nokia has used its opportunity as market leader to introduce an almost invisible, yet branded, Nokia language. It's important to note that this language does not necessarily draw new users, but Nokia's penetration of the market via traditional advertising has accelerated and even secured new purchases. Somehow the company has managed to overcome substantial manufacturing mistakes over the years, including everything from unexpected user errors to faulty displays.

Still, Nokia users keep returning to the Nokia brand. They return because people like the familiar. They're reluctant to change because they're essentially lazy and don't want to put in the effort that's required to master a new operating system.

As Nokia market penetration increases, and Nokia customers repurchase the brand again and again, this creates an ever greater loyalty that no traditional brand campaign can create. With each repurchase the Nokia language becomes further embedded into the behavior of the customer. This is priceless. And it all comes at a minimal cost because the Nokia brand language is out there capturing the unsuspecting consumer each time a phone rings within earshot.

## True Brand Power!

If you happen to be a Nokia phone owner, almost every element of the cell phone experience has turned into a branded Nokia experience. It's almost unnecessary for you to carry a phone charger because wherever you are, you're bound to find a Nokia charger that you can plug into—be it through a hotel or a friend. If you happen to own another brand of cell phone, you may not be quite so lucky. So something quite simple like the need to recharge has become part of Nokia's true brand power. A huge disadvantage for its current competitors.

---

### Highlights

We are all intimately familiar with our senses—if not always aware of them. They fully inform the picture of our daily life. When one of them is missing we realize how important it is.

However, for some reason the advertising industry communicates almost exclusively in a 2-D world. The fact is that a majority of the five thousand commercial messages we all are exposed to each day are based on what we see and hear—but only rarely on what we smell, touch, and taste.

Brand communication has reached a new frontier. In order to successfully conquer future horizons, brands will have to find ways to break the 2-D impasse and appeal to the three neglected senses. Superb picture quality won't do it. Rather we should look to embrace all five senses in order to create a foundation for future brand strategies.

Over the past decade the car industry has transformed every feature down to the very smell of the car into a branded exercise. Brands like Kellogg's, the breakfast cereal experts, no longer count on the natural crunching sounds of their product, but design these in sound labs. Singapore Airlines ensures that the aroma in the cabin is as consistent as the color scheme, which matches the makeup and uniforms worn by the hostesses.

Every detail of brands should be created with a true sensory signature. When formulating a sensory branding strategy, one of the key criteria for success is to design a platform that will allow you to carefully select the channels, the tools, and the senses you intend to tap into when building your future brand.

It is estimated that 40 percent of the world's Fortune 500 brands will include a sensory branding strategy in their marketing plan by the end of 2006. Quite simply, their survival will depend on it. If brands want to build and maintain future loyalty, they will have to establish a strategy that appeals to all our senses. This is a fact that no serious brand builder can ignore.

**Action Points**

- Determine to what degree your brand depends on our senses, either directly or indirectly. If possible, try to establish the nature of consumers' relationship with your brand.
- If the sensory appeal you've established in the first point were to be neutralized, and would no longer be of any value to your brand, what dollar amount would you then estimate to lose in terms of sales or brand loyalty?

- If the loss you've established above is considerable, what can you invest in it to maintain and enhance this unique brand asset?
- If neutralizing the sensory appeal of your brand won't affect the customer sales or loyalty—is this aspect being underutilized? Or is it simply impossible to leverage in a systematic way?
- Decide how you intend to handle this opportunity. Is it worth investing in? Before you do, read chapter 3, then make up your mind.

CHAPTER 3

# Smash Your Brand

**CARROTS ONCE CAME IN EVERY COLOR** *but* orange. There were red, black, green, white, and purple varieties. Then sometime in the sixteenth century Dutch growers decided to give this root vegetable a patriotic edge. Using a mutant seed from North Africa, breeders began developing an orange variety in honor of their monarch, William I, the Prince of Orange, who led them to independence against the Spaniards. A country with an orange flag now had an orange carrot. You might call this one of history's most superbly successful branding exercises, albeit one that was never capitalized on. Very few people who munch on a carrot—not even including Bugs Bunny—are aware that they're biting into one of the greatest missed branding opportunities of all time.

To place too great an emphasis on a brand's logo carries risks. Least of all there is a danger of neglecting all the other potential brand-building opportunities. If paid due attention, many other aspects may become recognizable in their own right. Color, navigation, texture, sound, shape. Even blindfolded, you'd know you're holding a classic Coke bottle. As you read earlier, in 1915 Earl R.

Dean of the Root Glass Company was given a brief to design a bottle which could be recognized by touch in the dark. And then, even if broken, a person could tell at first glance what it was.

Taking his inspiration from the pod of the cocoa bean, Dean produced a bottle with ridged contours. This led to the Coca-Cola Company's strategy to use the shape to emphasize the very brand.

## *It's Time to Kill Your Logo*

Remove your logo, and what do you have left? This is a very important question because a brand is so much bigger than its logo. Are the remaining components easily identifiable as yours? If not, it's time to smash your brand. The smash-your-brand philosophy considers every possible consumer touch point with a view to building or maintaining the image of the brand. The images, the sounds, the tactile feelings, and the text all need to become fully integrated components in the branding platform. Each aspect plays a role as vital as the logo itself.

## *Knowing What You're Known For*

Advertising messages are increasingly cluttering our airwaves and print media. The average consumer is bombarded with an astonishing 3,000 brand messages a day.[1] As each brand fights to be heard in this cacophony of the commercial world, it's vital that it strikes the perfect note to stand out. Our use of media itself has become more sporadic. As the broadcast media play on in the background of our busy lives, we have developed internal filtering systems that help us switch it off.

The consequent fragmentation of our attention requires advertising to develop a totally integrated brand message, optimizing every brand signal in such a way that the brand becomes instantly recognizable. This presents the advertiser with enormous challenges. As many as 20 percent of the tween generation (8–14-year-olds) own their own cell phones, and there's every indication that this number is increasing on an annual basis.[2] A large proportion of these phones are the basic variety, with noncolor screens, less sophisticated graphics, and minimal visual effects. How would your brand fare on this matchbox-sized canvas?

The situation of media fragmentation is further complicated by brand alliances. A large number—close to 56 percent—of all Fortune 500 companies have formed alliances in their communications. This fact alone demonstrates an even larger need for the implementation of a Smash Your Brand strategy, because when two brands share one space, and need to convey separate values and commercial messages in a limited thirty-second spot, two logos alone simply won't do the job.

Visit DualBook.com/bs/ch3/smash to learn more about brands pioneering the adoption of the Smash Your Brand philosophy in building and marketing their brands.

Only a few brands would pass the Smash Your Brand test today. Take a moment to consider your own brand. If you were to remove your logo and any other textual reference to your brand name, would your customers still recognize the product as yours? Chances are that without the logo and name, your brand loses its meaning. In order to reverse logo dependency, all other elements—colors, pictures, sound, design, and signage—must be fully integrated.

## *Smash Your Brand—Piece by Piece*

Smash Your Brand into many different pieces. Each piece should work independently of the others, although each is still essential in the process of establishing and maintaining a truly smashable brand. The synergies created across the pieces will be essential for your brand's success.

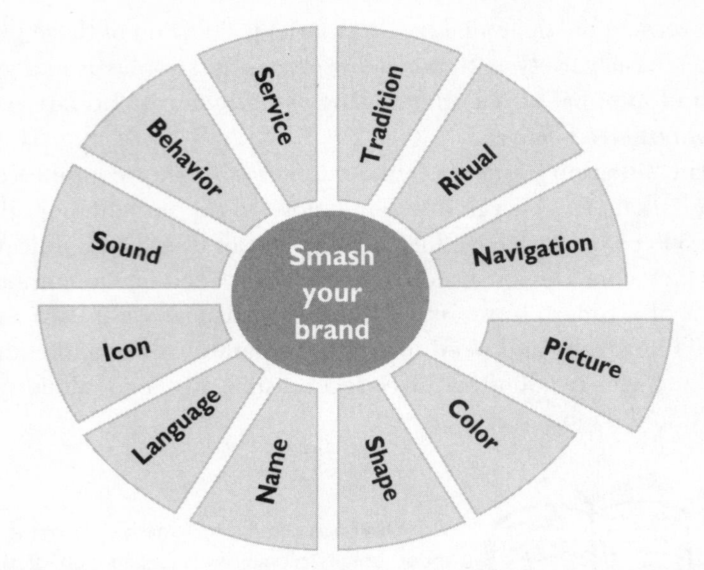


**FIGURE 3.1** Twelve components form the fundamentals of a Smash Your Brand philosophy.

## *Smash Your Picture*

Since its beginnings in 1965 under the name Benetton, the United Colors of Benetton has developed a consistent brand style identifiable in any size, in any country, and in any context. It was Benetton's intention to develop its own unique personality. They consider their clothing to be "An expression of our time." Their strategy in maintaining this integrity has been to generate all their own images. Luciano Benetton explains, "Communication should never be commissioned from outside the company, but conceived from within its heart."[3] Benetton is a brand that would survive smashing. The image and the design form its own statement and are part and parcel of the Benetton "heart."

Famous faces wearing white mustaches are instantly recognizable as the "Got Milk?" campaign, which has run for more than a decade. From the Williams sisters to Garfield, everyone who's anyone has done it. You can too! Just join Club Milk and post a picture of your milk mustache on the website. A white line across a lip is all you need to see to know you should drink milk because it's healthy!

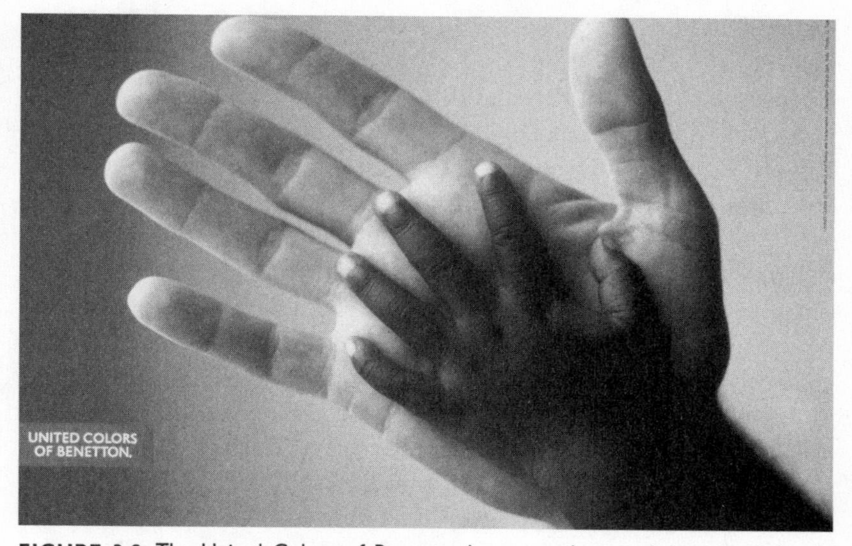


**FIGURE 3.2** The United Colors of Benetton has created a smashable pictorial style totally independent of the company's logo. *United Colors of Benetton and Sisley are trademarks of Benetton Group SpA, Italy. Photo: D.Toscani*

Too few companies would be able to pass this kind of brand identity test. Too many companies recycle images, or frequently change designers and photographers, and too often many different communication organizations are employed by the various departments in the same corporation. Packaging is done by X, marketing brochures by Y, and public-relations information by Z. This lack of synergy fragments the message, making the logo not only necessary but vital to distinguish the product. Corporate brochures are the worst offenders. Nonbranded stereotypical pictures of smiling people in suits around boardroom tables, a towering shot of the company headquarters, and the obligatory portrait of the CEO. One can't help but feel the waste of all this energy spent on publications that do nothing to build the brand.

## *Smash Your Color*

Coca-Cola has lost the battle for red in the European market. It had stiff competition from powerful local players. Thirty percent of the respondents *BRAND sense* questioned in the UK consider Vodafone the default owner of "Coke-red"; a smaller 22 percent still associate the color with Coke. Perhaps it would come as no surprise to see variations of the logo as part of Coca-Cola's marketing campaign in Britain. They're duplicating the classic red-and-white logo in different colors—including blue and green—to reflect the color scheme of each football team they sponsor. Green is also appearing in Europe and the Asia-Pacific region. In Germany, for example, the traditional red screw caps are now green. The trend follows through in the Japanese market, which also regards red as mostly belonging to others. Only in Coke's home market, the United States, does the brand remain most strongly associated with red. However, in a majority of the global markets, 36 percent of those surveyed do in fact associate red with Coca-Cola. A smaller 27 percent listed Vodafone (in markets where Vodafone is represented), followed by Budweiser and McDonald's taking a share of 13 and 12 percent, respectively, as the owner of red.

The major color advantage that Coca-Cola had over Pepsi in the cola war resulted in Pepsi turning to blue and establishing a global ownership of the color. In the *BRAND sense* study, 33 percent of the global population associates the color blue with the Pepsi brand. This would not come as good news to IBM, which for years was known as "Big Blue." In fact, results from the *BRAND sense* study confirm that in some countries, for example, Japan, people associate IBM with black rather than blue. Only 11 percent of today's consumers across major markets think IBM when they think blue. In fact, 14 percent stated that their color perception of IBM is unequivocally black.

## Fighting for Color Real Estate

A quick look at the logos of major corporations reveals that in color as in real estate, it's all about location, location, location. The result is an ever more frantic competition for the best neighborhood.

In 1942 Lucky Strike, the cigarette brand, struck a problem. The Second World War was raging and chromium, an element essential to the green ink on their labels, was in short supply. So at around the same time the American troops invaded North Africa, Lucky Strike released its new pack with its red target, along with the slogan "Lucky Strike has gone to war!" Six weeks later Lucky Strike sales were up 38 percent.

The primary colors clearly have dominated in the world of brands. However, there's no evidence to support the premise that red, blue, and yellow are more effective. It seems tradition is the only reason why these colors dominate. Over the past decades, many brands have tried capturing color ownership. Heinz, one of the world's leading manufacturers of quality condiments, launched their "Power of Red" campaign, which aimed to give women the confidence to wear red—and pour gallons of ketchup over the meals they serve. Steve McGowan, the senior brand manager, said, "Our packaging and our brand equity have been built over the years around the 'Lady In Red' concept, which has created a powerful connection between the feelings associated with the color red: energy, joy, control and confidence."[4]

Switzerland also lays claim to red. Switzerland holds the market on quality watches, knives, cheese, chocolates, and banks. Since the mid-nineteenth century it has conspicuously built its brands and leveraged its colors. When in 1863 the Red Cross was formed, it used the Swiss flag's colors in reverse and, albeit unintentionally, created one of the strongest brands in the world based on red and white. Any company qualified to say "Made in Switzerland" adds substantial value to its brand. Red and white have become synonymous with Switzerland, reflecting some of the most sophisticated state-merchandising exercise to date.

A British telecommunications company decided to position itself between red and yellow on the color wheel and launched Orange (the company name) in a campaign that proclaimed "The future's bright—the future's Orange." Part of their strategy was to offer students in major cities a free paint job. The only condition was that the color had to be, well, orange. Thus their whole campaign went a long way in laying claim to the ownership of the color orange. Is the telecommunications giant Orange a competitor for the airline company EasyJet? In the UK the color orange has become the center of a major legal dispute between these two brands, each one claiming the exclusive right to the use of orange.

Yellow is perceived to be the catchiest color of them all. At the beginning of the twentieth century, a telephone directory of services was launched and the Yellow Pages were born.

At around the same time a man by the name of John Hertz had a small stake in an automobile dealership in Chicago with a surplus in used cars. He hit on the idea of turning them into taxi cabs. At some point Hertz heard about a University of Chicago study that revealed yellow was the easiest color to see, and so he had these cars all painted yellow, and called the company Yellow Cab. When Hertz sold out, he went on to form Hertz Rent-a-Car and again made yellow the cornerstone of the new company's logo.

Transport and yellow seem to go hand in hand. Yellow and red are now being claimed by the global express courier company DHL. It's a popular combination that has seen McDonald's and Kodak wrestling for predominant ownership for decades.

When jewelry is presented in a robin's-egg-blue box, it takes on an added luster because the recipient knows this box is from Tiffany, the New York jeweler whose name has been synonymous with luxury, exclusivity, and authenticity since 1837. How much would a secondhand version of one of your packages sell for on auction? Some brands manage to impart magic and integrity through their packaging alone. Authentic Tiffany boxes and pouches have become marketable items, fetching up to $40 on auction sites. The larger the box, the higher the cost. Large boxes hold big items.

Tiffany's delicate blue forms the basis of the store's color scheme: it's the color of the catalogues, you see it in their ads, and of course on their shopping bags, too. No matter how much money you may offer Tiffany, you cannot buy a box from them. The ironclad rule of the company is that boxes (or pouches) leave the store only if they contain an item that's been purchased there.

To date the packaging of only a handful of exclusive brands can be found on auction sites. These include Louis Vuitton, Gucci, Rolex, and Hermès. This is a strong indication of a brand's ability to maintain its equity, but is also a fairly important indication of its smashability.

**FIGURE 3.3** Even a building under construction can pass the Smash Your Brand test. I don't think anyone would be in any doubt about the Louis Vuitton flagship store in Paris, which has recently undergone extensive renovations.

Can attention to color go too far? Try wearing an item of clothing from fashion giant Burberry when visiting pubs in central London, and you will find yourself at risk, precisely because British hooligans have adopted the Burberry colors as a form of identification. Rather than emanating luxury and class, they are a sign of a community that looks for trouble. As a result, in certain parts of the UK Burberry has suffered a distinct drop in sales. The power of smashable signals should not be underestimated.

Color is essential to the brand-building process because it's the most visible first point of communication. School buses, police cars, and garbage trucks are first and foremost distinguished by their color. Think mail vans, and the immediate thought is of their color. Using a color in a logo, and then sporadically splashing it across print materials, will not automatically build or maintain the color ownership. However, colors create clear associations, and it's these associations that will benefit your brand.

## *Smash Your Shape*

Shape is one of the most overlooked branding components, even though certain shapes clearly speak of their particular brand. Think of the bottle shapes of Coke, Galliano, or Chanel No. 5. Particular shapes have become synonymous with certain brands. The Golden Arches refer to McDonald's trademark, and they're consistently present at every outlet in every country all over the world.

Since 1981 the shape of the Absolut vodka bottle has been the primary component in every aspect of the brand and its communication. From fashion shows to ice hotels, footprints on the beach or northern lights, Absolut's inventive ads are all based on the shape of the bottle. The shape of the bottle is the shape of the brand.

You'd also recognize a Barbie doll and her body parts anywhere. She can be lying decapitated in a gutter and you would know her pert feet with high arches and long firm plastic thighs. Alternatively you may find just her head, and her big hair would be a dead giveaway. In other words, Barbie, in all her guises, smashes well.

Most computers look pretty generic, all except the iMac. No matter which generation you're referring to, you could smash it any which way and you would be left in no doubt that the fragments are part of an iMac. iMac would scream out its brand from the splinters of smooth plastic in bright transparent colors, or the bulbous "lamp" design with its movable flat screen. Even the shiny smooth earphones connected with the iPod would let you know that the shape of its shards is uniquely and ubiquitously Apple.

The curve of Barbie's waist, the graceful lines of the Apple, or the contours of the Coke bottle, each element that creates these products is fully integrated into their overall design, making the shape distinctly their own.

## *Smash Your Name*

When the Porsche 911 was introduced in Frankfurt in 1963 the model was called 901. The brochures were printed, the marketing material was all in place, and then everything had to be urgently changed. Much to Porsche's dismay, they discovered that Peugeot owned the rights to all three-digit model numbers of any combination with a zero in the middle, and this was nonnegotiable. Fortunately only thirteen vehicles got through the production line with the 901 insignia, and thereafter it became known as the 911.

Peugeot has held the numeric name rights for cars since 1963. The middle zero gives them a distinction that automatically identifies their models as Peugeot—even if you're not able to conjure up a mental picture of a 204 or a 504.

A similar strategy has been adopted by Absolut vodka. They deliberately misspell their brand extensions, using English words inspired by Swedish spelling—Absolut Vanilia, Mandrin, Peppar, or Kurant.

McDonald's uses the "Mc" in their name to every possible advantage. Their world is awash with Big Macs, McNuggets, McMuffins, and even McSundays. If you happen to receive an email from the corporation you'll be greeted with the words "Have a MACnificent Day." McDonald's naming philosophy is an essential part of their brand. This has resulted in many a legal battle like the one in Denmark in 1995, when McDonald's took Allan Bjerrum Pedersen to court for appropriating their name. He ran a small hot dog stand named McAllan. McDonald's was unsuccessful in this one. The claim against Pedersen was dismissed, and McDonald's was held liable for all the costs incurred.

Mac-ization of the language was formally recognized when Merriam-Webster added McJob to their collegiate dictionary, defining it as a low-paying job that requires little skill and provides not much opportunity for advancement.

The Disney Corporation has incorporated Disney characters into the structure of their Burbank, California headquarters. Twenty-foot-high statues of the Seven Dwarfs hold up the roof. Pathways take their names from other Disney luminaries—you can wander down Mickey Avenue and stroll across Dopey Drive. By using this naming strategy they leverage the brand, extending it to encompass every aspect of their environment.

The end result of these integrated naming strategies is that they reinforce the awareness of a brand's profile. The names are easy to remember. This further enables the company to concentrate their energies on other communication features rather than having to rebuild the brand again and again with each new product release. Subbrands become intuitively recognizable and tap into the broad set of values already well established by the parent brand.

## *Smashing Your Language*

Disney, Kellogg's, and Gillette are three completely different brands with one thing in common. Over the past decade they've established a branded language. The irony of this is that they may not even be aware of it. Whether coincidentally or purposefully, the *BRAND sense* study shows that 74 percent of today's consumers associate the word "crunch" with Kellogg's. Another 59 percent consider the word "masculine" and Gillette one and the same. Americans formed the strongest associations of masculinity and Gillette—by an astounding 84 percent.

There's one brand, however, that has scored higher in purloining language than any other. It is a brand that welcomes you to its kingdom of fantasy, dreams, promises and "magic." This will come as no surprise to anyone who has stayed at a Disney resort, taken a Disney cruise, or eaten in a Disney restaurant. It doesn't take long to hear "cast members" encouraging guests to "Have a *magical* day!"

Since the 1950s Disney has consistently built their brand on a foundation that's so much larger than their logo. A substantial chunk of the Disney brand relies on songs and voice-overs that always include Disney-branded words. Associating words with brands comes at no extra cost. And Disney has managed to "own" six of them:

> Welcome to our kingdom of *dreams*—the place where *creativity* and *fantasy* go hand in hand spreading *smiles* and *magic* at every *generation.*

The *BRAND sense* study shows that more than 80 percent—yes, 80 percent—of the world's population directly associates these generic words with Disney.

The key words are repeated over and over again in Disney's

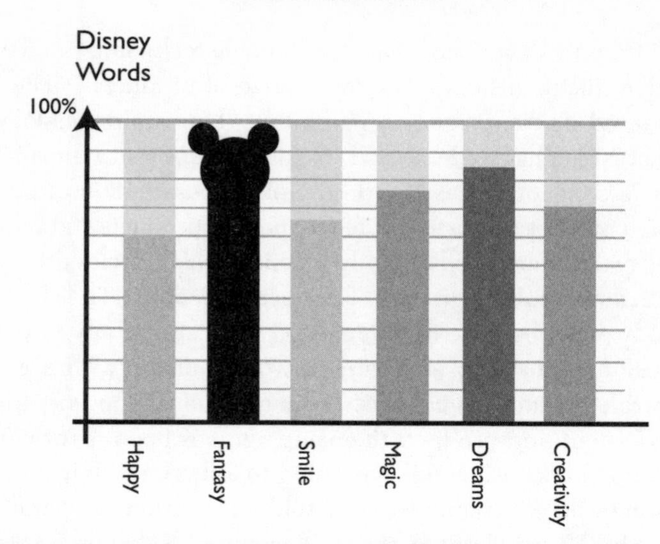


**FIGURE 3.4** Ninety percent of the world's consumers associate the word "fantasy" with Disney. Disney has created a truly smashable language that encompasses the administrative world of Disney, too. Disney doesn't have marketers, they have MarketEARS!

advertising copy, song lyrics, story lines, and on the Disney Channel. The words cross all media channels with ease and fluidity. No opportunity is wasted in making strong connections between Disney and Magic, Disney and Fantasy, Disney and Dreams, and so on. In the same manner that Orange, Coca-Cola, and the Yellow Pages have claimed their spot on the color spectrum, so Disney has succeeded in staking ownership of the language of fantasy, making it the place where magic happens and dreams come true. What's more, Disney's language survives the smash test. Pick a word, a sentence, or a column from any Disney publication, remove each brand reference, and *voilà* . . . the brand's still recognizable!

To create a truly smashable brand requires consistency and patience. This is a difficult requirement in a corporate world where the only constant is the ever-changing branding strategies and marketing presidents. Add to this a fluctuating financial market which demands instant results, and the brand message becomes just another bit of information in the overcrowded field.

For several years Nokia put out the word that its phones were "human" friendly. They say that it's "human technology and smart design [that] distinguish Nokia's wide range of products."[5] The company centered its campaigns on its unique "Human Technology"—a phrase that Nokia has trademarked. Nokia explains that Human Technology is "a concept that is based on Nokia's observation of people's lives, which inspires Nokia to create technology, products and solutions that meet real human needs."[6] Only recently Nokia decided to downplay this tagline, replacing it with "Connecting People." The choice has proved to be wise, because in the *BRAND sense* study, only 14 percent of the consumers surveyed associate the word "human" with Nokia.

In a strategy aimed to establish ownership of the notion of Nokia as a human-centered product, the campaign has been less than successful, largely because Nokia has failed to adhere to high standards of consistency across channels. They seldom mention the phrase and so nothing builds on or reinforces the concept of Nokia being the one and only provider of "Human Technology."

Nokia is not alone. Many companies have failed to convey their emotional strategy through the written word. For decades Colgate has talked about "Colgate Smiles," so much so that you would imagine that the word "smile" would be firmly in Colgate's corner. Well, not so. When it comes to owning the word association for "smile," Colgate ranks a distant third—following Disney and McDonald's. A closer examination of this supposed anomaly reveals an interesting phenomenon. The stronger the brand personality, the more human and less product-focused it is, and the easier the consumer finds it to associate words, phrases, and sentences with the brand.

Coca-Cola has used the word "enjoy" forever. It's on billboards, in ads, and even on their product label, yet Disney's characters have run off with 62 percent of the brand association, leaving Coca-Cola in second place with 53 percent. Likewise, McDonald's Ronald, M&M's animated candy mascots, and Kellogg's raft of characters are also popular associations with "enjoy." "Crunch," in contrast, has a unique association with Kellogg's.

The companies that take away word-association prizes have characteristically created detailed and fully realized characters that personalize their companies. These characters have almost become de facto spokespeople for the brand, lending it an engaging, human "voice." The point of the strategy is not to necessarily create characters, but to adopt a human-centered approach, and avoid product-centered tech talk that focuses on features.

This is far from the entire technique. It takes years for phrases, words, and sentences to be identified and accepted as "belonging" to specific brands. Communication has been built from the bottom up, not suddenly placed on top as a decorative bit of icing. The same message has existed since the inception of the product. The language which began with a few catch-phrases has been embraced and passed on from one generation of staff to the next in order to establish their own branded language. This language has then been consistently integrated across all channels, and just as you may know the sounds of French or Chinese, the words take on a sense of recognition.

The first step in integrating specific language into your brand is to identify the words you want to "own." Your selection should be based on those words you think best reflect your brand's personality. Choose words that are easy to integrate into many different kinds of sentences and which are the most flexible.

There's absolutely no mistaking Absolut vodka's language. Their Web home page[7] asks the Absolut Legal question: are you lawfully of age to drink? If you choose the "Yes" option, you are free to enter their world of Absolut Wonder. There you discover more about Absolut Reality. Should you wish to contact the company, click on Absolut Contact. Everything on this site is consistent with the Absolut advertising campaign—which has been running for over twenty years. It's a campaign that's based on continuity and variety, and 700 ads have been produced since 1980, all related to the original vision which launched Absolut Perfection.

The key to forming a smashable language is to integrate it into every single piece of communication that your company is responsible for, including all internal communications.

## *Smash Your Icon*

Icons or symbols are likely to become very important components in rebuilding your smashed brand. We currently operate in a world full of icons, and their numbers are clearly on the rise. Technology has given us many more channels opening up ever more varieties of advertising opportunities. Icons are a growing concern. They need to have an inbuilt flexibility to cross channels and they should be graphically sophisticated so that they can be equally understood on a billboard, computer screen, or cell phone display.

Icons have also been used in advertising to connect symbols, characters, and even animals with a brand. Just think about the Marlboro man, or Schweppes—a soft drink that has used bubbles as their trademark.

Successful icons help companies take their commercial message to new and unexplored terrains. Truly successful icons are also eminently smashable.

## *Smash Your Sound*

Brands the world over underestimate the value of sound. A while back I was sitting in a café. Over at the next table, someone's cell phone rang. The tone ringing was the well-known Coke melody, "Always Coca-Cola." In those short few seconds before the phone was answered, the tune had wormed itself into my brain, where it silently repeated itself the whole day. It sure says something about digital branding, because not only will the phone's owner hear this catchy Coke jingle several times a day, but anyone who happens to be in the vicinity will be exposed to the tune too.

Brands can be built using sound—not the sound that we take for granted on radio or television commercials, but more like background music that plays on websites, in stores, on hold buttons on the telephone, or even as ring tones. The Banyan Tree, the luxurious chain of resorts, hotels, and spas that specializes in tranquility and peace for the body and mind, plays the same subtle exotic music in their hotel lobbies as they do in their rooms. Furthermore, you will hear the distinct relaxing tones when you make your reservations on their website. This Banyan Tree theme is entirely smashable, in the same way the tunes played in the Mandarin Oriental and Peninsula

Hotels are. Each of these hotel groups, which operate extensively in Asia, have realized that music contributes as important a role in branding as the overall visual design does.

CNN and BBC World have consistently leveraged sound as their main brand feature. Does it work? According to BBC World it does. The television signature tune hit the best-selling music charts when the BBC released a special BBC World music compilation featuring all their compositions used as program and station themes.

Qantas, the Australian national airline, released a special music compilation of a children's choir singing "I Still Call Australia Home." The melody of this stirring song, which was originally written by an expatriate with emotional ties to his birthplace, was played on each Qantas plane during boarding and disembarking. It became part of every television and radio commercial, successfully generating a strong sense of emotional bonding between airline and consumer, the likes of which had not been seen before.

## *Smash Your Navigation*

You may be familiar with a supermarket chain, but unfamiliar with a particular store. Despite this lack of familiarity, you will probably still feel comfortable shopping there because the internal logic is consistent between outlets, and navigation follows more or less the same pathways. The canned-vegetable aisle is followed by the spice racks, which lead on to the pasta, where you'll also find your various tomato sauces. You can still pick up your gum at the checkout. If you blink you may even think you're back in your usual neighborhood supermarket. This is not happenstance, nor is it a case of déjà vu. Rather it's a carefully thought-out floor plan that's designed to meet your expectations of the store's brand.

Navigation—the way you find your way around a website, a department store, a supermarket—is entirely smashable. It also presents a challenge for companies to ensure the navigation remains consistent as their message crosses media channels. However, if we use the supermarket model, we know that the vegetables are just over from the dairy products, so even if there's a shelf of nuts in between, we are able to make the mental leap necessary. In the same way, there needs to be a synergy between your website, your cell phone campaigns, your store layout, your brochures, and your automated phone system, because they all link together.

Consistency is the only way to cut through the clutter of the noise. Navigation is one of the most essential tools that can be leveraged in building and maintaining this consistency.

The goal is simply to develop a smashable navigation, which reflects your brand. Whether the brand is a supermarket or merely an item on the shelf, there must be a consistency in the way you present it on every single channel you can think of. In order to do this you need to ask yourself three questions:

***1. Does the customer require a particular order in the way I present my product or service?***

In the same way that I feel comfortable with the order process at any Subway fast-food outlet across the world, so it is essential to establish a comfort zone for your brand whether it be online, offline, wireless, or a clicks & mortar combination. If you offer a free gift-wrapping

| Feature/Channel | Actions to be taken | Brochure | In-store | Website | Product |
|---|---|---|---|---|---|
| Navigation | Content structure | √ | √ | √ | |
| | Usage of Icons | √ | √ | √ | √ |
| | Color coding | √ | √ | √ | √ |
| Instructions | Tone | √ | √ | √ | √ |
| | Illustrations and icons | √ | √ | √ | √ |
| Feedback | Signal sounds | | | √ | √ |
| | Written or verbal status | | √ | √ | √ |
| Support | Support setup | √ | √ | √ | √ |
| | Handling of Q&A | √ | √ | √ | |

**FIGURE 3.5** Navigation has to be intuitive but it can also be branded. The object is to ensure consistency across all channels.

service in your store, then you should also offer it on your site. If your contact details are at the foot of each web page, then so they should appear at the foot of each catalogue page. If your special offers are always announced in red and yellow on a television commercial, then they should appear in red and yellow in print and on your website.

**2. *Should my customers be guided in their purchase choice by branded symbols or instructions?***

Some stores color code items into color categories. Cars make extensive use of icons on their dashboards. Some cell phones instruct the user as they go, rather than putting instructions into a separate manual. It's important to be aware that each of these components forms the characteristics of the brand. They become even more important when adding another channel to the brand, because it's vital that navigation should always be comfortable, easy, and familiar. That helps build loyalty, which is based on the brand's ability to continually communicate consistently, precisely, and in a branded context.

**3. *What perceived brand link does the consumer consider to be essential when viewing, using, or visiting my brand across channels?***

The links between channels are essential to identify and brand in a consistent way, making the usage—the process of finding or learning about the product or service—intuitive. When your navigation has reached that truly intuitive level, then you know that you have the foundation to make the brand truly smashable.

## *Smashing Your Behavior*

If you happen to visit the Animal Kingdom at Disney World, you will notice the service staff in the jungle next to the tigers speak in a thick New Delhi accent. In fact, every service component has been integrated to match the brand in the theme park.

Virgin has mastered this notion of consistency, too. Richard Branson leads the Virgin empire in conveying a sense of irony and humor in his casual, straightforward communication. The Virgin style, in turn, takes good-natured shots at established values. With a nudge and a wink to its audience, Virgin engenders goodwill and respect, and most importantly it makes the brand infinitely smashable.

At airport check-in counters, there are usually contraptions that indicate the maximum size of carry-on luggage. Airlines ordinarily go to great lengths to convey the legal and safety implications of their imposed restrictions. Virgin does this in its own inimitable way. In a friendly font it lets its passengers know that "You can have a huge ego, but only a bag this size (7kg limit)!"

The check-in line is equally painless. Smiling staff are on hand, the signage is super-friendly, and announcements are preceded by "Ladies and gentlemen, boys and girls . . . ," acknowledging passengers who are all too often overlooked. Even the Virgin flight is smashable, and the experience continues after arrival. Signs directing passengers to their oversize luggage humorously point out, "Size does matter!"

## *Smash Your Service*

If you're unhappy about any aspect of a product purchased at Harrods, the great London shopping institution, you can take it back, replace it, or simply get your money back. No problems whatsoever. The easy-return policy is just one area of service that Harrods has become famous for.

Smashing your service is as feasible as smashing all the other more tangible components shaping your brand. Passengers on Cathay Pacific receive a handwritten note from the staff wishing them a special journey. You may give a cynical groan and chalk it up to a standard script penned by a copywriter in the ad agency, but I was shocked to see that the passenger sitting alongside me received a similar note but with a completely different message.

The Peninsula Hotel in Chicago offers another brand of service. When I wished to listen to music in my room, I was told that this particular hotel had no CD library. However, I was asked to name whatever CD I desired and the concierge would happily arrange to have it delivered to my room, courtesy of the hotel.

Expectations vary depending on what a brand communicates to its audience and their individual perception of that message. Most companies overpromise and underdeliver. A rare few do the opposite. Louis Vuitton, maker of luxury leather goods, explicitly does not offer a lifetime warranty on its products. In fact, the company's documentation states a charge will be applied for repairs. The salesperson to whom you return your faulty product further reiterates this when you take it in for repair. But, based on my experience, when you come back to collect your item, you may well not have to pay for the service—and a salesperson will assure you this was done especially for you.

Visiting Amazon or eBay, you can always be sure your preferences have been stored, that their guarantee policy is consistent, and that you will receive relevant updates on products matching your profile. It has become an integrated service component of the online brand. The potential for overdelivery is everywhere. It's behind the promises your brand makes. It's about what consumers can do on your website. It's at home in e-newsletters. Your guarantee, stated response times, average waiting times, service statements all offer opportunities to underpromise and overdeliver. If I'm told you'll respond to my email within forty-eight hours, do it within twenty-four. If I'm told the warranty is for twelve months, don't deny service if I arrive with a faulty product twelve months and nine days after purchase. These extensions beyond the promise give you a perfect chance to offer the consumer more than he expected. Have you built in any unofficial policy that has the potential to thrill customers and woo their friends? It cultivates enormous goodwill, fosters positive word of mouth, and generates loyalty.

## *Smash Your Tradition*

When the debonair James Bond ordered his martini "shaken not stirred," the phrase took the martini to smashable heights. It's lingered on in cocktail parlance for forty years, and become a ritual.

The phrase is but one of the many smashable elements in the 007 spy series, and is heard in one James Bond movie after another. In each movie James Bond participates in a story line that has developed into a ritual all on its own. There are fast cars and always sexy women. The music is another essential element. Whether the song is performed by Shirley Bassey, Paul McCartney, or Madonna, the sound is first and foremost identifiable as Bond. It seems audiences around the world can't get enough of it. They wait for the next movie and flock to see it, knowing almost exactly what they're going to get.

The stronger the tradition, the more smashable it becomes.

Another wonderfully smashable theme is Christmas. From the tinsel to the Santas, the trees, fake snow, jingles, carols, roasting turkey, crackers and candles, and the red, gold, and green color combinations—almost every aspect of the season screams out Christmas. Along with tradition comes a swath of memories, and brands are often linked to the memories of traditional moments.

Brands need to understand in what context rituals appear, and where there is a potential to build in a branded ritual. Far too many brands ignore the importance of nurturing this phenomenon, overlooking a major opportunity to let the consumer "own" the brand and then become its ambassador.

## *Smash Your Rituals*

Can a ritual be trademarked? Apparently it can. Mars did just that when they applied for a trademark on the finger-scissors gesture they used in their Twix commercial. The Benelux Trademark Register accepted this, registered it, and now Mars is attempting to extend the trademark worldwide.

Evolving brand-based rituals are proving to be goldmines for brand owners. Take Nabisco's Mallomars. A biscuit "enrobed" in chocolate, it doesn't do very well in the heat. To avoid a Mallomar meltdown, Nabisco stops production from April to September. But as the weather gets cooler people start waiting for Mallomars to appear on the supermarket shelves. Erin Bondy, a spokewoman for Kraft, the parent company of Nabisco, says, "When they return, some media outlets told me they throw Mallomar parties."[8]

A similar tradition takes place in Denmark. At a specially designated date in November, a Christmas beer called Julebryg is delivered by horse-drawn wagon to selected bars in Copenhagen.

It is in the sports arena where we see rituals well embedded into the game. Before the national New Zealand rugby team, the All Blacks, play a game, they perform a Maori war dance known as the Haka. Maori people traditionally performed a Haka before charging into battle. On a less savory note, fans and players of Cardiff City Football Club in Wales have developed a ritual known as "Doing the Ayatollah." The club's fortunes have been less than sparkling, and after witnessing television footage of Iranians mourning the death of Ayatollah Khomeini, fans adopted the gesture of tapping the tops of their heads with the palms of both hands every time the team missed a goal or kicked outside of their boundaries. Whether performing the Haka or Doing the Ayatollah, these smashable rituals easily identify their teams.

Most rituals are generated by consumers. To date, few brands have seen the value in supporting consumer-generated rituals despite the enormous bonding that they can give rise to. Guinness drinkers are devoted to their black beer, but more than that, there's a ritualistic way of drinking it. Potential Guinness drinkers, take note that pouring the perfect glass of Guinness is an art form; it takes time.

- Chill the bottle or can for at least three hours. The gentlemen at Guinness suggest between 39°F and 45°F. Most find that a bit too chilly; 48°F to 52°F is preferred by most in the States.
- Start with a clean, dry 20 oz. tulip glass.
- Use the "two-part" pour. Pour the Guinness slowly into a glass tilted at 45 degrees, until it is three-quarters full.
- Allow the surge to settle before filling the glass completely to the top.
- Don't rush it, or the foam won't properly froth. The length of the total perfect pour should exceed two minutes; four to five minutes is better. You want the head to last for the whole beer.

Now that your perfect pint, with its creamy white head, is ready to drink, try creating a nice shamrock on the top of the head while pouring out the end of the foam.

- A 20 oz. Imperial Pint glass holds 20 oz. of liquid not including the head. It allows the regulation 18 oz. of beer, leaving 2 oz. for the head, plus what stands above the rim.

In contrast to most global brands, Guinness is one of the few that has established a raft of strong rituals around the actual consumption of the product. The brand is also closely tied to nationalistic feelings, as well as sporting institutions. Guinness's many rituals extend from how to order and drink the beer to how a sporting cheer can be recognized by the customers—blindfolded—without even mentioning the name. The brand has evolved from being a traditional beer brand—reflecting traditional brand loyalty—to a position where Guinness has fans rather than customers. The brand manages to fulfill each of the twelve components in the Smash Your Brand philosophy, including shape, color, language, and tradition.

## *Case Study: Smashing the Website*

Incorporating a smashable strategy is exactly what YellowPages.com created when it set out to reestablish its brand. Taking the Smash Your Brand philosophy into consideration, YellowPages.com made sure that every single element on their website passed the test. So whether online or offline, the consistency is there. YellowPages.com worked with a core consideration in mind. If their site was broken up into pieces, visitors should be in no doubt that whatever piece they were seeing was part of the YellowPages.com brand. Every element on the site is unique to the brand. If the YellowPages.com logo was not there, it would hardly be missed. The copy and the design speak of the brand without having to name it.

Each word has been carefully crafted to create a unique YellowPages.com voice that reflects the brand's overarching philosophies.

Even the ubiquitous privacy policy is written in easily understood language without compromising its legal obligations. Again, you can smash the copy and still hear the brand's voice in the broken phrases.

The site reflects the brand's offline function, going so far as to incorporate its failings (just as 1950s print technology didn't always achieve full registration, the site's colors don't always marry perfectly at the edges). Those minor flaws are used to create a comfortable, nostalgic relationship between visitors and the brand's service.

## Aging Gracefully

YellowPages.com is an example of a mature, considered use of the Internet, indicating the medium is approaching maturity. Yellow Pages brand builders use technology to communicate old-fashioned values, reflect offline simplicity, and convey humor that addresses visitors winningly. At last, the medium is led by reason and practical application rather than encouraging clichéd tech talk and whirring graphics. Technical virtuosity has often acted as an impediment to brand communicators wishing to convey their message with clarity and precision.

YellowPages.com is showing a mastery of the medium. Through applying humor, conveying core values, and consistently communicating a brand personality in every facet, the brand succeeds in taking on human qualities.

Branding means creating a brand personality. It's about making a brand human—and by doing so in a smashable way, you will be well on track.

---

### Highlights

Even blindfolded, you'd know you're holding a classic Coke bottle. And if that bottle were dropped and smashed, someone else could would be able to tell at first glance what it was.

Remove your logo, and what do you have left? A brand is so much bigger than its logo. Are the remaining components easily identifiable as yours? If not, it's time to smash your brand.

The Smash Your Brand philosophy considers every possible consumer touch point with a view to building or maintaining the image of the brand. The images, the sounds, the touch, and the text all need to become fully integrated components in the branding platform. Each aspect plays a role as vital as the logo itself.

Smash your brand into many different pieces. Each piece should work independently of the others, although each is still essential to establishing and maintaining a truly smashable brand. The synergies created across the pieces will be essential for your brand's success.

**1. Smash your picture**

Benetton is a brand that would survive smashing. The image and the design are its own statement and are part and parcel of the Benetton "heart."

**2. Smash your color**

A quick look at the logos of major food outlets reveals that colors create clear associations, and it's these associations that will benefit your brand.

**3. Smash your shape**

Think of the bottle shapes of Coke, Absolut, or Chanel No. 5. Particular shapes have become synonymous with certain brands.

**4. Smash your name**

McDonald's uses Mac or Mc in their naming strategy: Big Macs, McNuggets, McMuffins, McSundays. Their naming philosophy is an essential part of their brand. Subbrands become intuitively recognizable and tap into the broad set of values already well established by the parent brand.

**5. Smash your language**

Disney's language survives the smash test. Pick a word, a sentence, or a column from any Disney publication, remove each brand reference, and voilà . . . the brand's still recognizable!

**6. Smash your icon**

Technology has given us many more channels opening up ever more varieties of advertising opportunities. Icons are, as a result, a growing concern. They need to have a built-in flexibility to cross channels and be graphically sophisticated so that they can be equally understood on a billboard, computer screen, or cell phone display.

**7. Smash your sound**

Brands can be built using sound—not the sound that we take for granted on radio or television commercials, sounds such as the background music that plays on websites, in stores, on-hold buttons on the telephone, or even as ring tones.

**8. Smash your navigation**

Consistency is the only way to cut through the clutter of the noise. Navigation is one of the most essential tools that can be leveraged in building and maintaining this consistency.

**9. Smashing your behavior**

Richard Branson leads the Virgin empire with a sense of irony, humor, satire, and casual, straightforward communication. The Virgin style, in turn, takes good-natured shots at established values.

**10. Smash your service**

How would your customers characterize your service? Unique? Just as it is possible to smash all the other more tangible components, it should be just as possible to smash your service and still recognize your brand.

**11. Smash your tradition**

The stronger the tradition, the more smashable it becomes. Christmas is wonderfully smashable. So is James Bond. Movie audiences still anticipate hearing him order his martini "shaken not stirred," and the phrase took the martini to smashable heights.

**12. Smash your rituals**

Most rituals are generated by consumers. To date, few brands have seen the value in supporting consumer-generated rituals despite the enormous bonding that they can give rise to.

## Action Points

- ❖ If I smash your brand, remove your name, and take away your logo, would your customers still be able to recognize the brand?
- ❖ If your answer is yes, congratulations. Your goal is now to further enhance this unique proposition. If you are not one of the few whose brand has survived smashing, identify the three main reasons why it currently fails the test. Is it because of the lack of a consistent, prominent, and recognizable message, or is your brand the product of a brand structure that is too complex?
- ❖ Address each of the twelve components of the smashed brand. Score your performance in a diagram allowing space for two extra columns. Conduct the same experiment using your competitor, and then evaluate how you are doing in comparison.
- ❖ You are well on the way to enhancing the true value of your brand. Chapter 4 should inspire you to further work on developing a true sensory brand.

CHAPTER 4

# From 2-D to 5-D Branding

**UNTIL RECENTLY, MCDONALD'S** has not been coasting along as easily as they would have liked. Changing global food trends and lack of service were important factors in the decline in sales recorded in the latter half of 2003. Rethinking these aspects and addressing some of the problems associated with obesity have gone a long way to reverse the downward trend. The next challenge might be solving the simple fact that too many people still think Ronald smells!

A third of the consumers interviewed in the *BRAND sense* study thought McDonald's restaurants smelled like stale oil. In fact, more than a third of the consumers surveyed in the United States were quite harsh in their assessment, saying that their dislike of the smell puts them off their food—and off the brand. A larger 42 percent of those surveyed in Britain agreed. The United States and Britain are two of McDonald's largest markets. By comparison, Burger King fared better because only a third of the consumers in the United States, and 30 percent in Britain, felt the same way.

By the same token, the distinctive smell of McDonald's should not be underestimated. Paradoxically, half of the consumers say they love the smell of cooked food, and a visit to McDonald's makes their mouths water. However, Burger King again was consistently ahead of McDonald's in this group where 70 percent claimed a similar positive sensory association with the Burger King brand.

Food trends, worldwide, indicate that people are becoming increasingly health conscious. McDonald's has taken the initiative in developing healthier items for their standard menu—a sensory opportunity that they are not letting pass. To focus only on the smell would not paint the full picture of a brand under pressure. Quite a few people—14 percent—said the food looks unappetizing. And 15 percent were not happy with the aesthetics of the restaurants. Consumers in the UK were more damning. Some 24 percent of those affected by the noise in the restaurants said the sound in McDonald's gave them "negative" feelings. Upon further investigation of this negativity, it emerged that the sound of McDonald's is often equated with the sound of screaming kids, and in some instances with the electronic beep of the fryer timer.

## *Moving on from 2-D Branding*

To a large extent marketers have operated in a two-dimensional world, only occasionally venturing into a broader universe where they leverage all five senses. Increasingly consumers are expressing a more multidimensional desire incorporating a complete sensory approach. 4-D simulation games, which include sight, sound, touch, and smell are a permanent fixture at theme parks and video arcades the world over. In every major city of the world sweet aromas and heady scents waft out the doors of the many new shops specializing in scented soaps and candles, incense, fragrant potpourris, and aromatic oils. Aromatherapy in all its guises is designed to help create peaceful and relaxing environments.

As we age our senses become dulled. Some of our most powerful olfactory impressions were formed in childhood. Teenagers' sense of smell is 200 percent stronger than that of adults beyond middle age.[1]

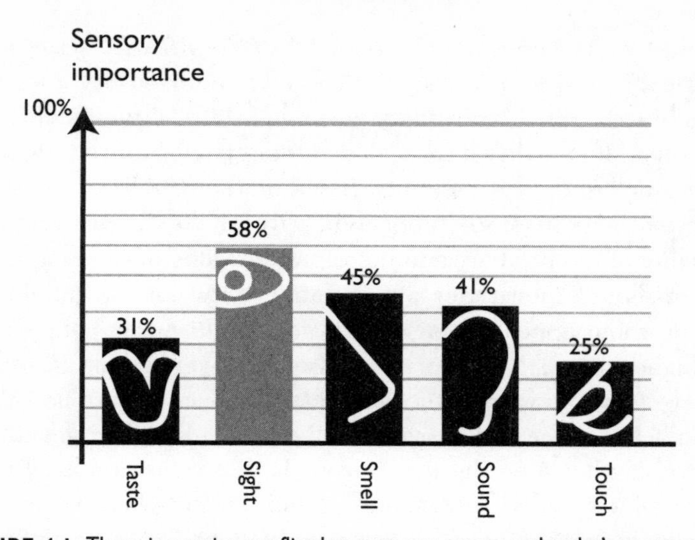


**FIGURE 4.1** There is a major conflict between our senses and today's communication. After sight, smell is the most important of our five senses. Source: Millward Brown and Martin Lindstrom.

Given the fact that children influence an astounding 80 percent of parents' purchases,[2] appealing to the sense of smell becomes increasingly important.

Of the sample surveyed in the *BRAND sense* study, 37 percent listed sight as the most important sense when evaluating our environment. This was followed by 23 percent listing smell. Touch ranked lowest on the scale. Generally, though, the statistics that emerged showed a very small differential when it came to a sense-by-sense evaluation, leading us to conclude that all five senses are important in any form of communication and life experience.

This conclusion does not come as a surprise. What is surprising is that the entire world of branding has ignored this for so long. Furthermore, what emerged in the *BRAND sense* study revealed that the more sensory touch points leveraged when building brands, the higher the number of sensory memories activated. The higher the number of sensory memories activated, the stronger the bonding between brand and consumer.

Almost every consumer interviewed in the *BRAND sense* focus groups expressed surprise at the lack of a multisensory appeal in today's brands. It's extraordinary when you think of the heights that brands like McDonald's have scaled without paying any attention to, say, the smell of their restaurants. Based on the *BRAND sense* study, we see that a multisensory appeal directly affects the perception of the quality of the product, and therefore the value of the brand. The study further demonstrates a correlation between the number of senses a brand appeals to and the price. Multisensory brands can carry higher prices than similar brands with fewer sensory features.

As we move away from the 2-D model, toward a more holistic 5-D approach, we need a Sensagram to help calculate the dynamics of the interplay between the five senses. It illustrates graphically the measure of a brand's performance, going beyond what we see and hear to include what we smell, taste, and touch. The Sensagram will give us a way to measure a brand's performance beyond what we see and hear, and include what we smell, taste and touch. Theoretically a very strong brand will appeal to all five senses, whereas weaker brands will appeal to only one or two senses.

The *BRAND sense* study also pointed to other variables that come into play. For example, the mention of a car brand could possibly

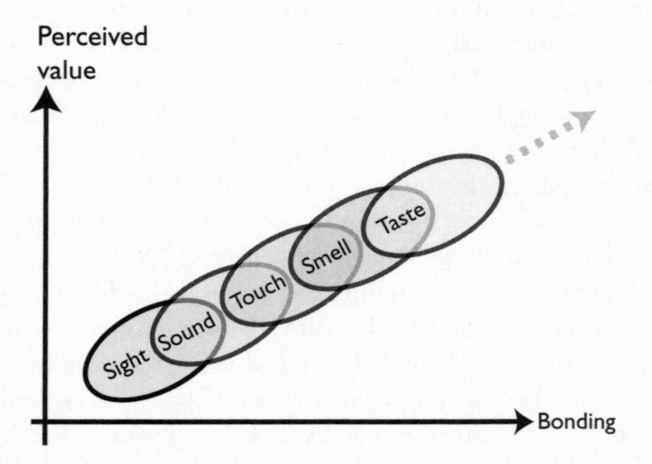


**FIGURE 4.2** The *BRAND sense* study shows a clear correlation between the number of senses a brand appeals to versus the price of the product.

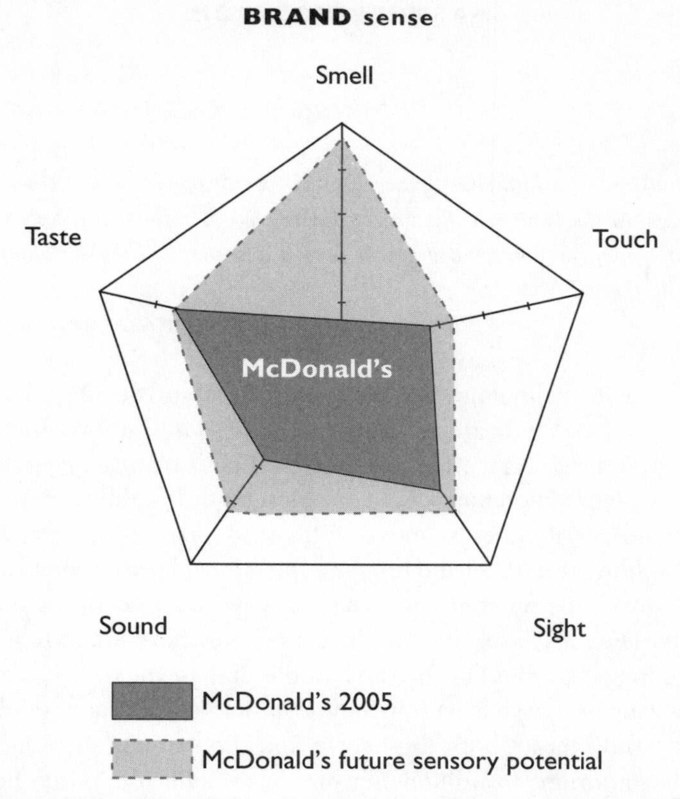


**FIGURE 4.3** McDonald's is the brand on the Fortune 100 list representing the biggest sensory conflict. The company is currently working to convey a healthier image.

evoke a sense of taste. This may not be related to anything other than the fact that many people eat in their cars. Some brands might conjure up negative sensory associations, like the smell in McDonald's. Such associations negatively impact on the total brand perception, and result in a low Sensagram score.

To explore the senses individually offers but one facet of the whole story at a time, making it almost impossible to create the full picture. Each sense is inherently interconnected with the others. We taste with our nose. We see with our fingers and hear with our eyes. However, just as we can identify a brand by a smashed bottle, so we can break down the senses to build up and generate positive synergies. With this holistic understanding, we can bravely enter the unexplored territory of sensory branding.

# *Sound*

*Sound puts us into the picture, or makes the picture more than an image. As the Inuit asks the visitor coming in out of the cold: "Speak so that I may see you. Add a voice, even a whisper, so that the other is really there."*

DAVID ROTHENBERG

When movie technology was new, people sat in theaters watching silent movies. The theater was never totally silent because these first moving pictures were often accompanied by a pianist playing along with the silent action happening on the screen. It's almost impossible to imagine a contemporary movie without sound. Sound is fundamental to building the mood and creating the atmosphere of whatever celluloid story's being told. Sound is hard-wired into our emotional circuits. The muscles in the middle ear of a newborn infant reflexively tighten in preparation for the pitch of the human voice.

Hearing is passive and listening is active. The sound of your brand should target both the hearer and the listener since both are equally important in influencing purchase behavior. While hearing involves receiving auditory information through the ears, listening relies on the capacity to filter, selectively focus, remember, and respond to sound. We use our ears to hear and our brains to listen. Sound is emotionally direct and so it should be considered a powerful tool.

The way a brand sounds should never be underestimated, because it can often be the deciding factor in a consumer's choice. More than 40 percent of consumers believe the cell phone sound—that is, the sound of its ring—is more important than the phone's design.

In a study published in the *Journal of Consumer Research*, Ronald E. Millman demonstrated that the pace of music playing in the background affected service, spending, and traffic flow in stores and restaurants.[3] The slower the music, the more people shop. The faster the tempo, the less they spend. Related studies have shown significantly longer dining times for restaurant tables when slow music was played. This resulted in more money being spent at the bar. The average bill for diners was 29 percent higher with slow music than with fast.

Even if we're more involved in hearing than listening, our mood is still affected by what we hear. In a study undertaken by Judy Alpert and Mark Alpert, which looked at how music affected mood, they concluded that happy music produces happy moods.[4] However, sad music resulted in greater levels of purchase intent, lending credibility to the saying "When the going gets tough, the tough go shopping."

## The Power of Sound

A fascinating experiment took place in a village on the Mornington Peninsula in Australia. Local residents, alarmed by the increase in street crime, got together and decided that the best way to tackle the problem was to remove the offenders from the main street after nightfall. Instead of taking the usual more-police, greater-security, and tough-on-crime stance, they chose to play classical music. On every block the sounds of Mozart and Bach, Beethoven and Brahms could be heard. In less than one week, the town reported a dramatic decrease in crime. It was so successful that the main train station in Copenhagen, Denmark adopted the same approach, as did the Port Authority Bus Terminal in New York City.

The Bellagio Hotel and Casino in Las Vegas experienced first-hand the power of sound. They took special note of the buzz of slot machines and the shower of coins falling into the winner's tray. Great sounds to the winner's ear, but supposedly disheartening to the neighbor still pulling that handle and getting nothing but the whir of a losing combination. For a while the hotel replaced its noisy slot machines with "cashless" ones, but to their dismay, they found their slot-machine revenue noticeably falling. It seems that a machine's not a slot machine unless it whirs and jingles—and this applies to losers and winners alike. In no time at all, the original machines were restored to service.

David Anders, a gaming analyst for Merrill Lynch, agreed with the move. He says the tourist market "is not ready" for coinless slots. The sound of coins popping into and flowing out of the machines is part of any casino's ambience. The sound "generates excitement and calls attention to the area. It lets people know other people are winning. With cashless slots, I guess you'd hear the buzz of the printer."[5]

## Staging a Branded Mood

Music makes new memories, evokes the past, and instantaneously can transport you to another place. All three characteristics are present in the Disney World universe. Carefully choreographed sound is piped through the entire park. Even the bird sounds are controlled. It's a whole environment designed to capture the hearts of children and waken the child within each adult. Theme music and recognizable tunes sung by well-loved and well-recognized characters are an essential part of the total Disney World experience. From the razzmatazz at the main gate to the up-tempo walking music playing in the streets, the music helps monitor your mood.

As retailers struggle to find ways to differentiate their stores from their competitors', so they're beginning to integrate multisensory components. NikeTown, Borders, and Victoria's Secret are among a steadily growing list of companies leveraging more than just sight and sound. Victoria's Secret plays classical music in their stores. It creates an exclusive atmosphere and lends an air of prestige to their merchandise. These companies are not alone. Today global audio branding company Muzak has an audience of 100 million people listening to branded tunes every day across malls, stores to elevators.

Sound and sight are the two senses that are already integrated in every aspect of marketing and merchandising. Traditionally, sound has focused on appealing to our hearing, at the expense of our listening capabilities. The notion that sound can actually influence a purchasing decision has been pretty much ignored.

Sound is becoming more sophisticated, and you will first need to evaluate what role sound will play in your product or service. Specific sounds are associated with specific goods—and sometimes we're not even aware of it. Obviously businesses that are based on audio will focus absolutely on sound. Where sound is an important component of the product, it should be leveraged to support it. Products that are not dependent on sound can use it as an adjunct to the product. No sound should be ignored.

Because consumers are surrounded by a low level of white noise from washing machines, dishwashers, blenders, coffee makers, toasters, and the like, manufacturers opted for no sound at all. What they found was that by removing the sound, products tended to lose part of their "personality." Strange as it may sound, they also lost a means of communication with the consumer. Companies have learned this the hard way. In the 1970s IBM released its new improved model 6750 typewriter. The beauty of it, so they believed, was that they'd managed to make it a silent operator. Typists did not like it. They simply could not tell if the machine was working or not. So IBM added an electronic sound to reproduce the functional noise they'd worked so hard to eliminate!

Good audio design has begun emerging in other industries, demonstrating that sound can add something extra to a brand. Luxury cues are often subconscious. Take a car door. How inclined would you be to buy a car whose doors closed with a hollow tinny sound? The way the door closes is more important than you imagine. In the mid-twentieth century when the Japanese were seeking to produce a high-quality car, they formed the first unit whose sole responsibility was to manage the challenge of a "branded car sound."

We need look no further than the Japanese-designed Acura TSX to see just how sophisticated car manufacturers have become—particularly in the area of sensory branding. The engineers methodically refined the design of the door sashes to reduce high-frequency resonance when the doors shut They also designed a special "bumping door seal" that purposefully transmits a low-frequency vibration to the door itself, creating a sound of "quality."

Nearly a third of consumers surveyed for the *BRAND sense* study claim they can distinguish one car brand from another by the sound of their doors closing. Japanese and Americans consumers are the most sensitive to this phenomenon, with 36 percent in Japan and 28 percent in the United States agreeing. Only 14 percent of consumers across the countries surveyed cannot distinguish a difference.

Sound is given special attention by car makers and it's no surprise that before a product hits the production line, its sound has been created by a multidisciplinary team consisting of sound engineers, product designers, and psychologists who ensure that the sound of the product will enhance the values and convey standards of trust, safety, and luxury that befit the brand.

Attention to the quality of sound is now spreading across a wide range of industries. Toy companies, computer hardware, kitchen appliances, and electronic goods—manufacturers are adopting standard sound quality monitoring and are now conscious of characteristics of sharpness, loudness, tonality, roughness, and fluctuation. A specific sound will add another point of differentiation to their brand.

## The Branded Sound of Bentley

In July 2003 one of the world's most prestigious cars, the Bentley Continental GT, was launched. This was a £500 million project.

One of the main aims in car acoustics is to reduce noise. Noise from the wind, the road, the suspension and the engine should not intrude. The interior should be supremely comfortable, overall noise reduced to a blissful hum, offering the ultimate in driving pleasure. This new model not only had to look like a Bentley, but it had to sound like one. From the very beginning acoustic engineers decided how the car should sound, and only then did they begin working to achieve this. This was such a fundamental consideration that the engineers were able to influence the design of the car, ensuring that both the intake and exhaust manifolds made a true, unique, and instantly identifiable sound.

Bentley carried out extensive research among existing Bentley owners, testing new additions to the brand as well as benchmarking the sound quality of other luxury sports cars. They ended up with a sound for the Continental GT that is deep, smooth, muscular, and inspiring. A smart move in a market where 44 percent of the consumers indicate that the sound of a car is an important factor in their choice of the brand.

## Once Upon a Sound

The focus on branded sounds is far from new. In 1965 a famous call was registered. It was broken down into a series of ten sounds alternating between low chesty and high falsetto. Defined in a ten-point description ranging from the first, "semi-long sound in the chest register" to the last "long sound down an octave plus a fifth from the preceding sound," this yell belongs to Tarzan. No one can copy it without due acknowledgment.

The power of Tarzan's call, the chimes of the NBC network, and the well-known MGM lion roar are sounds millions have been familiar with for decades. And then came Microsoft's start-up sound for Windows. Windows is the operating platform for 97 percent of the worlds PC users, meaning that more than 400 million people listen to a Microsoft start-up sound every day.[6]

Did Microsoft leverage the opportunity? According to the *BRAND sense* study they did, but only partly. Across major markets, 62 percent of the consumers surveyed who have access to a Windows operating system with speakers recognize Microsoft's start-up sound, and associate it directly with the Microsoft brand. Consumers in the United States and Japan proved to be the most au fait, in contrast with the European customers, who were substantially less familiar with the sound.

Given the fact that this is a sound that so very many computer users hear daily, the numbers that recognize it are relatively low. This becomes understandable when you look at the history of the Windows sound. Since its first release in 1995, Windows has changed its sound four times. The original three-second start-up tune was composed by an avant-garde musician, Brian Eno. His brief was to create a sound that would be inspiring, universal, optimistic, futuristic, sentimental, and emotional. Microsoft has missed an opportunity to build on their enormous market share. There is a lack of consistency of sound across all Microsoft channels spanning software, PDAs, phones, games, television, and the Internet. Until very recently Microsoft has overlooked the power of sound, leaving the corporation in a situation where a substantial effort needs to be dedicated to what could potentially become one of the strongest branded tunes in commercial history. If they continue to change the start-up sound, it will take on a generic character in the public's mind, and next to no one will recognize it.

## Nokia's Secret Weapon

The bars of musical notation in Figure 4.4 will not mean much at first glance, but it's these simple notes that have given Nokia a considerable competitive advantage. This is the music for the Nokia ring tone, and it's been trademarked.

Nokia is the world's largest cell phone manufacturer, and so that tune is played, and heard, millions of times a day all around the world. This amounts to thousands of hours of branded sounds for each individual.

Over the years Nokia has spent a considerable amount of money marketing the company. But they've hardly spent a cent on promoting its tune—also known as the Nokia tune. Despite this fact, the tune is recognized the world over. Let's do the math. On average, a cell phone rings around nine times a day. The average length of the ring is about eight seconds, leaving a person exposed to more than seven hours of ring tones a year! And this without even considering the considerable amount of ring-tone sound heard by other people bypassing cell phone holders. So, how well has Nokia managed to harness this major branding opportunity?

Slowly but surely Nokia has built a significant awareness of their brand just by leveraging something as simple as their ring tone. The

**FIGURE 4.4** The Nokia tune played by millions of consumers . . . every day! *Artist's rendition of Nokia tune*

*BRAND sense* study shows that 41 percent of the consumers across the world recognize and associate the tune with the brand when hearing a Nokia cell phone ring. In the UK the number is considerably higher, with 74 percent recognizing the tune, whereas in the United States, there's 46 percent recognition. Perhaps it's no coincidence that the most repeated melody in the movie *Love Actually,* starring Hugh Grant. is the Nokia tune. Filmed in London, the storyline has integrated the Nokia tune as an important ingredient when the insecure Sarah, played by Laura Linney, reveals her addictive relationship with a Nokia mobile phone. No surprise the Nokia tune has become an integrated sound phenomenon in the UK.

## The Secret Language of Nokia

Nokia has created an extraordinarily high sound recognition across the globe. This sound register involves an almost subliminal recognition associated with various cell-phone functions: accept, reject, recharge, flat battery, and even an inbuilt alarm to wake you or remind you of an important appointment. Chances are you're so familiar with the sound palette that you can recognize the sound language of Nokia without even being aware that you know it.

Close to half of the consumers who recognize the Nokia tune associate it with very positive feelings. It appears that a Nokia phone doesn't just ring. More than 20 percent who hear the Nokia name mentioned says it makes them feel positive—generally "pleased," "excited," "satisfied," "cheerful," or "in control." This is a branding tool that taps directly into the emotions. Today every Nokia cell phone contains the Nokia tune when leaving the factory, and perhaps because not all of us are as technical as we should be, more than 20 percent never change the default ring tone. In those cases where the consumer changes the ring tone, 86 percent just pick a tune from the menu created by the cell phone manufacturer rather than downloading a different tune from a website.

Based on their 39 percent market share, let's assume Nokia has produced some 400 million cell phones. And assuming that all these phones are still in use, 80 million people are currently listening to more than seven hours of the Nokia tune a year. This number might be lower because it does not take into account the very many consumers who choose other ring tones as well as the fact that every Nokia phone leaving the factory today is installed with a default Nokia start-up tune as well.

Over the past five years Nokia has established a solid indirect branding machine which feeds on our senses in a frequent and very effective way. The astonishing fact remains that Nokia is not paying one single dollar to secure such enormous exposure.

## Motorola's Search for the Right Tune . . .

Besides Microsoft, Nokia's closest competitor is Motorola. Motorola is struggling to achieve the same sort of brand awareness that Nokia enjoys. However, an astounding 11 percent of the consumers surveyed in the *BRAND sense* study mistook the Motorola ring for a Nokia tune. The percentage is even higher in the United States, Motorola's home market, where 15 percent confused the brands, generally assuming, however, that the tune was Nokia's. Only 10 percent of the consumers worldwide recognize the Motorola ring, and 13 percent in their home market. But these are early days. There are still many sound opportunities that are ripe for exploration.

## Intel versus Nokia

It wasn't that long ago if you mentioned the word "microprocessor" you were likely to get a mystified stare. Few mainstream consumers knew anything about the processor, even though it was the "brain" that powered the computer. But today many personal computer users can recite the specifications and speed of the processor, in the same way that car owners can tell you if they have a V4, V6, or V8 engine. The awareness of "Intel" has grown along with the awareness of the chip.

Launched in 1991, the Intel Inside program created history. It was the first time a PC component manufacturer successfully communicated directly to computer buyers. Today, the Intel Inside program is one of the world's largest cooperative marketing programs, supported by thousands of PC makers who are licensed to use the Intel Inside logos and spend close to $200 million a year on marketing on top of the cooperative marketing program, rumored to have a

$1 billion price tag. Did Intel get value for their money? Most definitely, when 56 percent of consumers across the developed world recognize the Intel Inside tune. Yet it is intriguing to realize that Nokia achieved sound awareness with only a limited investment, while Intel spent millions attempting to achieve the same thing. This makes it the only product in the world no one has seen, heard, or touched; yet by using sound and vision as the main pillars of their branding strategy, people the world over can dance to the Intel tune.

### It's Not About Nokia—It's About You

Every product has a sound. Your Siemens microwave pings; your Miele dishwasher ding-dongs; your BMW's doors, your Dell computer, and your Seiko wristwatch all have their characteristic sound. Nonelectronic sounds also pervade our life. Corks pop. We hear the opening of the milk carton, the crunching of our cornflakes, the bubbling of a freshly poured soda. There are thousands of brands that have yet to realize the enormous potential available by tapping into sounds and making them a feature of the brand.

One thing is certain: It's only a matter of time before your competitors start making some noise.

## *Case Study: Bang & Olufsen—Branding the Sound of Falling Aluminum!*

Is there a similarity between the sound of aluminum tubes falling onto cobblestones on a street in Denmark and the sound of the ring of a corded phone? A strange juxtaposition? Perhaps not. If I were to sit you down in a room full of traditional telephones, each ringing the ring of a branded phone, would you be able to identify which brand is what? Naturally you would hear the difference from phone to phone, but would you be able to pick the brand?

Whether we're talking about AT&T or GE, Panasonic or Sony, no corded phone manufacturer has designed a distinct, user-friendly, and branded sound similar to Nokia's cell phone tune—except for one which in 1993 released their latest model and broke the branded silence.

When the Danish luxury hi-fi manufacturer Bang & Olufsen commissioned composer and musician Kenneth Knudsen to design a unique, smooth, attention-demanding sound for the next corded BeoCom2 telephone, the challenge was to think laterally and come up with a sound that was recognizably distinct. This sound not only had to be distinct but would also serve as an unmistakable audio logo branding Bang & Olufsen.

The result is evident. Knudsen combined the sounds of falling aluminum tubes with musical notes, a sound he believes reflects the whole concept of the BeoCom2. He says, "We call it a ringing tune rather than a tone, as it contains many more elements than a simple note. This ringing tune has an acoustic texture of metal and glass, representing the physical components of the phone itself. Within one second, we wish to communicate a mood, a feeling, an impression; just like you get when you meet the physical product." The BeoCom 2 ring tune has raised the bar in the manufacture of corded phones. By refining this existing sensory touch point, additional brand equity is established, and another aspect of the Bang & Olufsen brand enters the universe.

Furthermore, the difference between the BeoCom2 sound and other ringing tones is that the Bang & Olufsen sound is human, it makes you feel at home, and it's instantly recognizable, says Birgitte Rode, CEO of Audio Management. Poul Praestgaard, Senior Technology & Innovation Manager at Acoustics Research, states that the "humanization" element will become standard in all future developments from Bang & Olufsen. This attitude perfectly supports the core value of the brand, and adds another sensory dimension to the identification of the product.

## *Sight*

*The question is not what you look at, but what you see.*

HENRY DAVID THOREAU

The human brain updates images quicker than we see. It accommodates every turn of the head, every movement, every color, and every image. In describing vision, Dr. Diane Szaflarski says, "The efficiency and completeness of your eyes and brain is unparalleled in comparison with any piece of apparatus or instrumentation ever invented."[7]

Vision is the most powerful of our five senses. Understandably, it is the sense that brand builders and marketers have traditionally concentrated on. According to Geoff Crook, the head of the sensory design research lab at Central Saint Martin's College of Art and Design in London, 83 percent of the information people retain has been received visually. He goes on to say that this is probably because they don't have other options.[8] The question is if this fact is still relevant. Every indication from the *BRAND sense* study suggests that of all the senses, smell is the most persuasive.

The reality is that copious quantities of visual information bombard us all the time. Clutter sets in, and so the visual effects fail to pack the big punch that they could. Their power is dissipated by the sheer volume of visual stimuli.

Only a small 19 percent of consumers worldwide believe the look of an item of clothing is more important than how it feels. Whereas a good half of them place the emphasis on feel rather than appearance.

The fashion industry is not alone in experiencing this swing in preference from look to feel. The food industry is seeing a similar, although less dramatic, pattern emerge. More than 20 percent of consumers say that the smell of food is more important than the taste. Rather than assuming this to be a rejection of design or long-standing taste preferences, it is an indication of the emergence of the other senses taking their place in the holistic scheme of a sensual universe. Whatever the case, there's no escaping the fact that distinctive design generates distinctive brands, and successful brands are by their very nature visually smashable.

Nothing is immune. Any man who's come to rely on his little blue, diamond-shaped pill will know it anywhere. It's his Viagra. Viagra is one of the jewels in the pharmaceutical company Pfizer's crown. Former presidential hopeful Senator Bob Dole became one of the paid public faces of this drug for erectile dysfunction. In the ads he refers to the pill as his "little blue friend." He promises it "changes lives for the better."

Viagra is an excellent example of how color and shape can be used effectively, and be protected by a trademark. This combination of pharmaceutical brand identity and product design is globally recognized. By leveraging the visual components of the tablet, Pfizer has helped Viagra secure trademarked brand loyalty beyond the time when their patent expires.

Pharmaceutical companies usually distinguish their product by color and shape. Accudose tablets which treat thyroid conditions come shaped like a thyroid gland. On television ads AstraZeneca promotes their leading anti-ulcer medication as the "purple pill." This is not new. Over thirty years ago the Rolling Stones referred to a 5mg Valium in their song "Mother's Little Helper." They sang about women who needed the extra help of little yellow pills to make it through their stressful days.

Tablets and capsules come in all shapes, sizes, and colors, each intended to differentiate the product, impart a particular emotional "feel" to the drug, and instill customer loyalty. The way a tablet looks is an important aspect in maintaining loyalty. When AstraZeneca decided to replace Prilosec with Nexium, they used the same color and called it the "*new* purple pill."

## Shaping Brands

Little blue pills are one way of changing individual lives, whereas innovative use of shape can change a whole city's life. Take the recession-plagued town of Bilbao in the Basque region of Spain. This industrial port had a dream to revitalize their image. They wanted to reinvent themselves. After years of planning and negotiation with the Solomon R. Guggenheim Foundation, they hired Frank O. Gehry, an innovative architect, to build a museum unlike any other on a large site in the middle of the town.

Gehry designed an organic sculpture. Its swooping titanium-clad curves house a museum that's so spectacular that the Bilbao Guggenheim has become one of Europe's most popular new sites. People are flocking there just to experience the Guggenheim's galleries. Bilbao, once just another tired industrial town on the European map, has been transformed by a building, which beckons with its courageous, daring, and totally unique shapes.

Innovative architectural structures often become iconic trademarks instantly synonymous with the cities where they are. Only

Sydney can claim the billowing sails of the Opera House that sparkle on the foreshore of its harbor. Jorn Utzon's revolutionary design with its organic shapes and lack of surface decoration adds to Sydney in every way—it's a venue for performing art, people congregate on its broad stairs, street performers line the walkways, and it offers some of Sydney's most spectacular views. The Sydney Opera House and Guggenheim Bilbao are both totally smashable.

Shape is an instantly recognizable visual aspect of any brand. When Theodore Tobler designed a triangular shape for his chocolate bar, the shape stood out more prominently than its taste. In 1906 it was against the law for chocolate makers to use their Swiss heritage in their logo, so as a way of proclaiming his nationhood, Tobler used the Matterhorn mountain to inspire the shape of his product. Fearing that a competitor would copy his concept, he applied for a patent on the manufacturing process in Bern. This was granted and Toblerone became the first chocolate product in the word to be patented.

Seventeen years after Theodore Tobler patented his chocolate, Milton S. Hershey registered his Hershey's Kisses, and turned his plume-wrapped chocolates into a cultural icon. Over the past century an entire Hershey world has been built around the original Hershey's Kisses foundation stone. Every day 25 million Kisses roll off the production line in Hershey, Pennsylvania. It's a town that bills itself as "The sweetest place on earth." The place is "built on chocolate." Streetlights are shaped like Hershey's Kisses, and there's accommodation, amenities, and activities "in all flavors"![9]

Hershey Park is one of the main attractions: entertainment day and night. Food halls that serve up Hershey chocolate milkshakes and Hershey's Kisses brownies. You can hold conferences at the Hershey Lodge, stay at Hotel Hershey, and pamper yourself at the spa with treatments like Whipped Cocoa baths and Chocolate Fondue wraps. Sweet indulgence indeed.

Chocolates aside, many products have based their identity on their distinct shape. The liquor industry has been at the forefront. Take the distinctive Galliano bottle, which is shaped like a classical Roman column. Finlandia vodka, Kahlúa, Bombay Gin, Johnny Walker, and Hennessy XO cognac are all products whose bottles' very shapes define their brand.

More recently the liquor industry has looked to the perfume and fashion world for inspiration. Coco Chanel loved perfume bottles. She even displayed the empty ones on her vanity. She said, "Those bottles are my memories of surrender and conquest . . . my crown jewels of love." She believed that "the bottle is the physical manifestation of the scent it contains, daring, seductive, alluring."[10]

The liquor industry wants to emulate some of the hope and promise that perfume bottles communicate. This is not as far-fetched as it sounds. Packages hold mystery and intrigue. Statistics show that 40 percent of all perfume purchase decisions are based on the design of the bottle. Jean-Paul Gaultier has taken this notion all the way with Fragile, his perfume for women. Fragile comes in a brown cardboard box with the word "Fragile" stamped on it in red. Inside the intriguing package is a magical snowball. Shake it up and a thousand golden flakes dance around a Fragile woman. Close to two million bottles have been sold.

The auto industry is another industry where shape plays a vital role. In many car models shape has become the defining feature. Think Beetle, Mini, and the military-inspired Hummer. Within this distinguished shape-centered crowd, the Lamborghini has carved its own special niche because it's one of the few vehicles where doors open upward instead of outward. This unique feature is trademarked. No one else can manufacture a car with this type of door.

Distinctive shapes create the most solid foundation for brand building across channels. We recognize and remember shape, and this may be what accounts for the longevity of Hershey's Kisses, Toblerone, and the Beetle.

## *Touch*

*Joy has a texture.*

OPRAH WINFREY

Wine bottles are now coming with screw tops rather than corks. Does wine sealed with a cork taste better than wine sealed with a screw top? Probably not, at least not for any recent vintage. Now, this is undeniably a matter of perception, but irrational as it may be, I imagine the wine that I've just poured from an uncorked bottle is superior. The screw top reminds me of soda, and just fails to assure me of the quality of the wine. The tactile sensations associated with opening a bottle of corked wine are lost.

How a brand feels has a lot to do with what sort of quality we attribute to the product. People still go around kicking the tires of a car they're thinking of buying. This may have been a reasonable test of quality many years ago, but today it's as irrational as the concept of the cork somehow adding to the taste of the wine. However irrational it may be, the feel of a product is essential in forming the perception we have of the brand.

The way a car feels when we sit inside it and run our hands over the steering and controls is important to 49 percent of consumers making a choice. Less than 4 percent of the people surveyed suggested that the tactile feeling of a car is irrelevant.

Britain's Asda supermarket chain, which is a subsidiary of Wal-Mart Inc., cottoned on to the economic advantages of touch. They removed the wrappers from several brands of toilet paper so that shoppers could feel and compare textures. This has resulted in soaring sales for its home brand, and an extra 50 percent store space has been allocated to their product.

To counteract the Florida heat, Disney World sprinkled chilled water on people hovering outside its shops, luring them into the air-conditioned world of merchandising.

Tactile qualities of a brand are often not quite as obvious as this bottom line. Perhaps one of the most intriguing results that emerged in the *BRAND sense* study occurred in the cell phone industry. One would imagine that cell phones that allow consumers to customize their look and ring would be permanently vulnerable to the ever-changing parade of newer, more fashionable models. Apparently not. The results revealed that 35 percent of the consumers interviewed stated that the feel of their phone was more important than the look. An astounding 46 percent of U.S. consumers said that the weight of the phone was more important than the look of it in their purchasing decision.

As electronic goods are manufactured in smaller and lighter versions, so consumers' perception keeps apace. Although heavier

| TOUCH (%) CATEGORY | Home Ent | Sports wear | Fast food | Soft drinks | Auto | Phone | Ice cream | Soap |
|---|---|---|---|---|---|---|---|---|
| Some/very important | 11.6 | 82.2 | 10.4 | 15.1 | 49.1 | 43.9 | 21.7 | 61.5 |

**FIGURE 4.5** Since the early 1900s, the Coca-Cola Company has taught the consumer to associate soft drinks with touch. Is this about to change due to Coke's revised strategy? The Coca-Cola Company was the first soft drink company in the world to teach the consumer to associate soft drinks with touch. Today, more than 100 years later, 15 percent of consumers still have strong tactile associations with the soft drink category.

weight intuitively feels more substantial, we like the convenience of small and light. There is one important proviso, however, and that is that the device in question is made from quality materials. We don't want our digital cameras to feel like plastic, nor do we want our PDAs to feel like tin. We demand superior craftsmanship with the most innovative materials.

Many electronic goods are going retro. A newer range of digital cameras are taking their inspiration from pre-digital cameras because consumers are demanding more than technology for its own sake. The average size of a man's hand is just too large to feel comfortable and confident manipulating a camera that can fit snugly in his palm. Then there's the issue of shutter sound. These new-generation larger cameras have added an artificial shutter sound that lets you know that the picture's been taken.

Something as simple as a remote control can tell you a whole lot about the quality of a brand. The heavier the remote, the better the quality—at least according to the consumer, who makes her quality evaluation on the feel rather than the look. This might explain why one manufacturer of luxury high-fidelity equipment, Bang & Olufsen, added weight to their remote control. They have consistently sought to build the ultimate quality items, but have worked equally hard to ensure that consumers have the best possible perception of their brand. They've carefully studied every aspect of their engineering, from the weight of the remote to the way their products open and close to the precision sound generated by the micro engines.

## Losing the Grip on Coke

In 1996 the Coca-Cola Company worked on Project Can, which was designed to transform most of their packaging from bottle to can. By the end of 2000, Coca-Cola had the first prototype ready for production. The famous bottle was about to morph into a bottle-shaped aluminum can. But an unexpected hitch developed. The new shape of the full can was unable to carry the weight when stacked. There was no accounting for the amount of damage from crushed cans and spillage that would result. The project stalled and was eventually canned. Coke cans were doomed to share their shape with every other soda on the market, with only red to distinguish the brand.

Did the Coca-Cola Company abandon their shaped-can project prematurely? A year later, Sapporo Breweries in Japan managed to achieve what Coke had been working on for years. They released the world's first bottle-shaped can. The part-bottle, part-can container developed by Daiwa was an instant success. The distinct taste of Sapporo beer together with the unique shape of the can proved a winning combination.

Although the *BRAND sense* focus groups confirmed that Coke still ruled supreme in shape recognition in countries that sell Coke in a bottle, this is the first time the company has lost the edge to another brand. Over the years, Coca-Cola has seen an erosion in the distinct characteristics that were once uniquely associated with their brand.

A STEADY DOWNGRADE When the Coca-Cola glass bottle was introduced with its distinct shape, size, and weight, it became an instant icon. As the company embraced new technology and adopted plastic bottles and aluminum cans, the tactile sensations that were so strongly associated with their product steadily eroded. A blindfolded person could not tell if the can he was holding was Coke or any other brand of soda.

The dissipation of the brand did not stop at the can. As sales of post mix increased—that's when the drink is mixed from syrup and carbonated water—and the brand was no longer served in any recognizable container, it failed to even be recognized as anything other than cola. Furthermore, in order to secure massive distribution through restaurant-chain outlets, the company agreed to have its product served in paper cups marked with the restaurant's logo. The only way you know you're getting a Coke is if you see the logo on the dispensing machine, because you drink it out of a cup marked McDonald's, or Burger King, or Wimpy, or KFC!

Even though limited statistics are available on the number of plastic versus glass bottles sold, cans and post mix, the trend is clear. Today it is estimated that more than 98 percent of all Coke sold in the United States is served in plastic, metal, or paper—but not glass. The fact is that most consumers in the United States who want to drink their Coke from a classic bottle have to make a concerted effort to seek it out.

According to the *BRAND sense* study, 59 percent of the world's consumers prefer their Coke in a glass bottle. This includes 61 percent of the U.S.-based customers and 63 percent of those in the UK. Despite this evidence, the company continues to cut its production of Coke in a bottle, diminishing one of its most important assets. These statistics from the *BRAND sense* study confirm that the distinct Coca-Cola touch is slipping through the company's fingers. What emerges from the study is that in those countries where the glass bottle has been replaced with plastic, Coke's tactile advantage has been displaced.

Coke is suffering. Our global study shows that their major com-

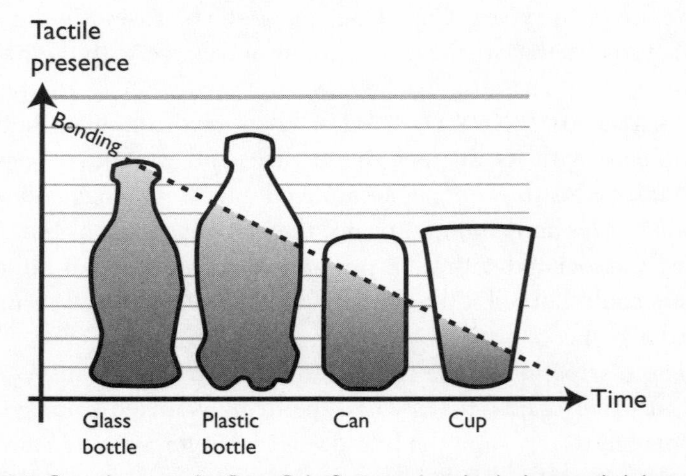


**FIGURE 4.6** Over the years the Coca-Cola Company has slowly downgraded the strong tactile signature of the brand.

petitor, Pepsi, is gaining an edge on touch. Within the home market, 60 percent of American consumers state that Pepsi represents the stronger tactile sensation. In contrast, only 55 percent of the consumers surveyed believe that Coca-Cola was distinct. Even though some statistical uncertainty has to be taken into consideration, that's an amazing five percentage points less than their century-old enemy. A similar picture appears when asking U.S. consumers how the old rivals fare in the physical sensation department. Coke leads by a marginal 6 percent, scoring 46 percent over Pepsi's 40 percent.

**IS COKE PREPARED FOR THE FINAL BATTLE?** In the international marketplace glass-bottle Coke is very much alive and still kicking in the face of a well-planned replacement process. Wherever the bottle is still sold, Coke emerges as the clear tactile leader in the soft-drink market. In Europe 58 percent of the consumers in our study stated that they still perceive a unique tactile feeling when drinking a Coca-Cola, in contrast to Pepsi's 54 percent.

A similar yet closer battle is taking place in Japan, where the majority of drinks come in glass bottles. Through our group sessions across the world we noticed that consumers in Spain, Poland, Britain, Denmark, South Africa, Germany, India, and Thailand were all able to describe the tactile feeling of the glass Coke bottle. This unique touch point no longer exists in the United States. The irony is that this loss could have been so easily avoided. Sadly, this failure is likely to be repeated when Coke's bottle-replacement plan gears up in the international marketplace.

This is not only a cautionary tale about how Coca-Cola is losing its tactile grip, it's also the story of how economic efficiencies in production and distribution have consistently downgraded the look, sound, and feel of the product. In addition to this, the difficulty of maintaining quality in the mix machines across the globe has further weakened the distinctive taste of the product. Several hundred undercover staff members from The Coca-Cola Company daily visit hundreds of bars ensuring that the bar staff doesn't mislead the customer who orders a Coke in good faith. All this results in a consistently weaker brand across not three but all four senses. This downgrade is proving to be lethal to the brand.

## *Smell*

*Smell is a potent wizard that transports us across thousands of miles and all the years we have lived.*

HELEN KELLER

The smell of a rose, a freshly cut lawn, mothballs, vinegar, peppermint, sawdust, clay, lavender, freshly baked cookies . . . Our olfactory system is able to identify an endless list of smells that surround us daily. Scents evoke images, sensations, memories, and associations. Smell affects us substantially more than we're aware of. We underestimate just how large a role it plays in our well-being. Smell is also processed by the oldest part of our brain. It's played a vital role in our survival, alerting us to distant danger, like fire. Animals communicate through smell. Through smell they instinctively know to reproduce, how to find their prey, and how to avoid danger.

Smell can alter our mood. Test results have showed a 40 percent improvement in our mood when exposed to a pleasant fragrance—particularly if the fragrance taps into a happy memory.[11]

There are about 100,000 odors in the world—a thousand of them being classified as primary odors, and then there are a myriad combinations of multiple odors. Each primary odor has the potential to influence mood and behavior. Everyone perceives odor differently, since so very many other factors come into play. There's age, race, and gender, to name but a few intervening variables.

Our smell preferences have changed over time. In a study conducted by Dr. Alan Hirsch, the neurological director of the Smell and Taste Treatment and Research Foundation in Chicago, consumers were asked to identify smells that brought on moments of nostalgia. What he discovered was that there was a divide between those born before 1930 and those born after. Those born before 1930 cited natural smells, for example, pine, hay, horses, and meadows. Those born after 1930 were more likely to mention artificial smells like Play-Doh, marker pens, and baby powder. The year 1960 proved to be another watershed for the smell of freshly cut grass. Those born before liked it, those born after associated it with the "unpleasant necessity of having to mow the lawn."[12]

In chapter 2 we established that everyone likes the way a new car smells, but what emerged in the *BRAND sense* study is that some cultures are more affected by smells than others. A huge 86 percent of consumers in the United States find the smell of a new car appealing, whereas only 69 percent of Europeans feel the same way. Branding cars has moved beyond stylish design and powerful engines in order to make the car a multisensory experience.

## Eau de Rolls-Royce

Hundreds of thousands of dollars have been spent reproducing the distinct smell of the 1965 Silver Cloud Rolls-Royce. The smell is impossible to buy! Yet it has been an essential component in maintaining one of the world's primary luxury brands. It's a small masterpiece in sensory branding.

When Rolls-Royce started getting complaints about their new models not quite living up to their illustrious predecessors, they worked out that the only difference between the new models and their older ones—apart from the obvious—was the smell.

The interiors of older "Rollers" smelled of natural substances like wood, leather, Hessian, and wool. Modern safety regulations and building techniques mean that most of these materials are no longer used, and have been replaced by foams and plastics. The only way to recapture that essence was to artificially mimic it. Using a 1965 Silver Cloud as a reference, the team began a detailed analysis of its aroma, identifying individual odors. They formulated a chemical blueprint for the essence of their analysis. In total, eight hundred separate elements were found. Some, like mahogany and leather, were expected, but others like oil, petrol, underseal, and felt were more surprising.

With this analysis in hand they manufactured the smell. Now, before each new Rolls-Royce leaves the factory, the unique smell of Rolls-Royce is added to the undersides of the car's seats to re-create the small of a classic "Roller."

Essentially this Rolls-Royce story illustrates the importance of maintaining perception—often without being aware of what the perception actually is.

## Eau de Car

Cadillac works equally as hard as Rolls-Royce to ensure loyal customers. General Motors is making sure that nothing a potential buyer touches, hears, or smells is left to chance. Cadillac's new-car smell, that ethereal scent of factory freshness, is a result of customized engineering. In 2003 they introduced a special scent, which is processed into their leather seats. The scent—sort of sweet, sort of subliminal—was created in a lab, chosen selected by focus groups, and is now a part of every new Cadillac put on the road. It even has a name. It's called Nuance.

For years the leather used in luxury cars has been tanned, processed, and colored in order to neutralize its natural smell. It was then injected with industrial aromas. Today a process called "retanning" puts fragrant oils back into the leather. Research shows that our preferences have changed. We now prefer the smell of artificial leather to real leather. Car manufacturers are now going to great efforts to satisfy customer demands, and they have created an artificial leather smell for our cars.

This takes branding to a whole new level. Ford has a specific branded aroma, which they've used since 2000. Like Ford, Chrysler uses a single fragrance for all their cars. Other manufacturers use different scents for different models. This marketing strategy bears fruit. According to the *BRAND sense* study, 27 percent of U.S. consumers believe Ford vehicles have a distinct smell, although only 22 percent say the same about Toyota. An even more dramatic trend occurs in Europe, where 34 percent consider the Ford smells distinct in contrast to only 23 percent who say the same about Toyota.

Ever since Dr H. A. Roth performed his simple yet powerful

color-and-flavor tests in 1988,[13] companies have been trying to develop tools to ensure a synergy between consumer perceptions and sensory realities. Symrise, one of the world's leading flavor and fragrance companies, working with experts from international universities, has developed what they believe is the way to achieve sensory synergy. Called Organoleptic Design, this technique incorporates flavor and aroma as a fundamental part of the design process, and will ensure a synergy between what consumers taste and smell with what they see, touch, and hear.

Organoleptic Design does away with traditional packaging, which denotes flavors simply by varying the color of the package. For example, this technique requires far more than yellow color to differentiate a lemon-flavored pack of gum from, say, a peppermint one. The process begins when a panel of users are presented with a "blind" package. There are no outward markings or indications of what flavor exists within the plain wrapping. The panel is asked to taste or smell the product. Based on their comments, panel members are then asked to develop a flavor or fragrance imagination map illustrating how they perceived the taste/aroma experience.

The purpose of this creative task is to interpret the different flavor or fragrance "worlds" in initial packaging or product design ideas. Extensive fine tuning of the prototypes aims to get closer and closer to representing the "real" consumer experience. By using the Organoleptic Design tool, Symrise can create a sensory synergy between product, brand, and sensory experience that hits the right note with the consumer.

## *Taste . . . and Smell*

*Smell and taste are in fact but a single composite sense, whose laboratory is the mouth and its chimney the nose . . .*

JEAN-ANTHELME BRILLAT-SAVARIN

Smell and taste are known as the chemical senses since both are able to sample the environment. They are closely interlinked. Many studies indicate that we often eat with our nose—if food passes the smell test, it will most likely pass the taste test. In the *BRAND sense* survey, when questioned about the smell and taste of McDonald's food, consumers tended to react positively to both smell and taste, or negatively to both. They didn't hate the smell but loved the food, or vice versa.

It is possible to leverage aroma without including taste. However, taste without smell is virtually impossible. Taste is closely related to smell, but it's also closely related to color and shape. Look no further than the language of chefs who talk about retaining color, natural color, and deep color. We associate certain colors with certain tastes: red and orange are sweet, green and yellow are sour, white is salty.[14]

The use of taste to support products is by its very nature extremely limited. Despite this, there are still unexplored opportunities that could be exploited. Even the most obvious "taste" products—for example, in the dental care business—have so far failed to leverage this opportunity. Take toothpaste. It has smell, taste, and texture, yet it's rare to see this leveraged across other product categories. The smell and the taste of major toothpaste brands could be extended to encompass dental floss, toothbrushes, and toothpicks. Currently the only synergy we see in this department is, with few exceptions, the use of the brand name and the corporate colors.

Apart from the obvious physical barriers that limit leveraging taste, philosopher and author Susan Sontag describes the elusive nature of this sense: "Taste has no system and no proofs." Smell works over long distances. Taste does not. Our emotions can be triggered by a vague whiff of a bygone smell. A mothball can conjure up warm and cuddly feelings for a grandparent, or the smell of motor oil can take you back to when you were Dad's young helper fixing the family car.

These bygone associations are referred to as the Proust phenomenon, and are named after Marcel Proust, the great French novelist famous for his memoirs in the early twentieth century. The Proust phenomenon is increasingly being triggered by branded smells. In older studies a large group (80 percent men and 90 percent women) reported having vivid, odor-evoked memories that trigger emotional responses. In 1987, *National Geographic* surveyed 1.5 million readers and questioned them about six odors. A. N. Gilbert and W. J. Wysocki reported on a subgroup of 26,000 within the survey. Half of those who were forty years of age and over could connect a memory to at least one of the six odors. Memories were recalled in response to both pleasant and unpleasant odors, particular if the odors were intense and familiar. Dr. Trygg Engen of Brown University conducted studies that contradict earlier findings about the predominance of vision, and concludes that our ability to recognize scents and odors is much greater than our ability to recall what we have seen.[15]

Smell, touch, and taste are important in the language of love. To touch and taste another taps into our most elemental selves, and so the species continues. In fact, it's been shown that extracts from male sweat can affect the regularity of a woman's menstrual cycle.

Pieter Aarts and J. Stephan Jellinek are psychologists who have studied how people's feelings, judgments, and behavior are subconsciously shaped by odor. They refer to this as the Implicit Odor Memory.[16] Their findings support the premise that fragrance is a factor when someone buys, collects, or uses a product. We can therefore conclude that odor plays a very important role in consumers' acceptance of a brand. Increasingly aroma is becoming a substantially effective brand vehicle. Visual power has become dissipated in a world that bombards consumers with all kinds of visuals. There's so much visual clutter that people are becoming skilled at moving through it wearing "blinkers." Given this overexposure, attention to visual messages has decreased.

Two identical pairs of Nike running shoes were placed in two separate, but identical, rooms. One room was infused with a mixed floral scent. The other wasn't. Test subjects inspected the shoes in each room and then answered a questionnaire. Consumers overwhelmingly—by a margin of 84 percent—preferred the shoes displayed in the room with the fragrance. Additionally, the consumers estimated the value of the "scented" shoes on average to be $10.33 higher than the pair in the unscented room.[17]

Another experiment was conducted in Harrah's, a casino in Las Vegas. One area was set aside and infused with a pleasant odor. Over the next few weekends, the revenue of the machines was compared to the earnings of the machines in the unscented zone. Revenues from the scented area were 45 percent higher than those from the scentless counterparts. Understandably, over the past few years Harrah's casino has spent thousands of dollars to see whether fresher air, wider aisles, and back supports can increase gambling—today most casinos in Las Vegas, including Bellagio, The Venetian, and Mandalay Bay, have implemented similar strategies.

The Hilton in Las Vegas even went so far as to release a scent manufactured by Alan Hirsch, a Chicago neurologist. The scent, called Odorant 1, was placed in a slot machine pit, and the increase in revenue paralleled that in the Harrah's experiment.

The capacity of a brand to include aroma as part of a sensory experience naturally depends on what type of business it is. But whatever the line of business, a steady increase in branded smells is taking off as we speak.

## Smell and the Supermarket

All around the world people and companies are becoming aware of the power of scent. A Disney World popcorn attendant has a wonderful working knowledge of how smell affects his business. He knows that when business is a little slow, all he has to do is turn on his artificial popcorn smell and in no time at all he has a line waiting for his popcorn. Woolworth's in Britain knows this too. In a buildup to the festive season, twenty of its stores introduced the smell of mulled wine and Christmas dinner. W. H. Smith, the largest newspaper/magazine chain in Europe, also went all out for Christmas and introduced the smell of pine needles.

Victoria's Secret has their own blend of potpourri, giving their lingerie an instantly recognizable scent. Superdrug used a chocolate odor in a Central London store on Valentine's Day. The London Underground filled some of its busier platforms with a refreshing perfumed scent called Madeline. They hoped it would add a bit of cheer for its three million passengers, not to mention offering them a moment's break from some of their less hygienic fellow commuters.

Several chains stores are starting to introduce branded smells. Thomas Pink, a British store that specializes in fine shirts, has introduced sensors in their stores that emit a smell of freshly laundered cotton to passing trade.

The future of brand building is about not only inventing new sensory appeals, but also identifying the brand's existing sensory assets.

These assets can be trademarked and leveraged across any new brand extensions. Technology is making identification of odors a viable option, as well as creating the formulas required for brands to "own" their own scent. Crayola is one of the many companies that has begun seeking to trademark its most distinct smells, starting with their crayons, their primary product, which have no doubt left their odor imprint on the memories of millions of children who drew with them.

From a situation driven more or less by a series of coincidences, every move by the global flavor and fragramce industry indicates a change toward more brand-focused behavior. Over the years the brands have developed sensory trademarks, but now they are also letting the brand drive the development process.

The flavor and fragrance company Symrise has smelled the direction. Instead of following the normal aroma and flavor development processes, they have decided to let the brand become the centerpiece of its own development. The move indicates the beginning of a manufacturing trend in the flavor and fragrance industry.

By turning the very scientist-driven development process upside down, companies like Symrise now ensure that the brand values will control every signal sent form the products.

In the big picture, each sense can be leveraged to build a better, stronger, and more durable brand. This cannot be done in isolation. The object is to ensure a positive synergy across a multiple of consumer touch points. And each of these branded customer touch points can be trademarked, ensuring a unique identity that will be impossible for any competitors to copy. The road ahead is not necessarily easy. There are many challenges awaiting. Sensory signatures that characterize the brand need to be identified, and it's vital that the consumer feels comfortable with your sensory brand. It's far from an easy task—but it's possible.

---

## Highlights

To a large extent marketers have operated in a 2-D world, only occasionally venturing into a broader universe where they leverage all five senses. Increasingly consumers are expressing a more multidimensional desire incorporating a complete sensory approach. Of the sample surveyed in the *BRAND sense* study, 37 percent listed sight as the most important sense when evaluating our environment. This was followed by 23 percent listing smell. Touch ranked lowest on the scale. More generally, the statistics show a very small differential when it came to a sense-by-sense evaluation, leading us to conclude that all five senses are extremely important in any form of communication and life experience.

### Sound

Hearing is passive and listening is active. The sound of your brand should target both the hearer and the listener, since each is important in influencing purchase behavior. While hearing involves receiving auditory information through the ears, listening relies on the capacity to filter, selectively focus, remember, and respond to sound. Many elements of our everday life are clearly associated with sounds. If we don't hear them, we miss them. The sound of a brand adds to the perception of product quality and function. If removed, the perception is diluted. It is therefore extremely important to assess the role of product-generated sound because, increasingly, consumers are becoming more aware—and critical—of this phenomenon.

### Sight

Vision has until recently been perceived as being the most powerful of our five senses; however, research indicates that this no longer is true. Whatever the case, there's no escaping the fact that distinctive design generates distinctive brands, and successful brands are by their very nature visually smashable. Tablets and capsules come in all shapes, sizes and colors, each intended to differentiate the product, impart a particular emotional "feel" to the drug, and instill customer loyalty. The automobile industry is another category where shape plays a vital role. In many models shape has become the defining feature.

### Taste and Smell

Smell and taste are known as the chemical senses since both are able to sample the environment. They are closely interlinked. Smells affects us substantially more than we're aware of. Test results have showed a 40 percent improvement in our mood when we're exposed to a pleasant fragrance—particularly if the fragrance taps into a happy memory.

---

## Action Points

Is your brand ready to take on a leadership role as a true sensory brand within your category?

- ❖ Identify the three most interesting ideas you learned from reading this chapter.
- ❖ To what degree would your brand be able to leverage one or several of these ideas?
- ❖ Write down the top concerns you might have implementing these ideas.
- ❖ Enough inspiration. Chapter 5 will help you to execute all the great ideas you've identified and evaluated so far. Read on.

CHAPTER 5

# Stimulate, Enhance, and Bond: Crafting a Sensory Brand

**WHEN THE NATIONAL AUSTRALIAN BANK,** one of the world's largest banks, released its first-generation website it took close to a full minute to open. Like many other banks, the National was known for long lines at the tellers. An independent survey was conducted asking visitors' general impressions of the new website. Even though there was no direct relationship between the long lines at the bank and the long download time of the website, consumers saw the website as just another waiting experience caused by the bank. The survey elicited comments to the effect of how the bank had managed to do in cyberspace what it already did in its branches—keep people waiting.

Perception of a brand is as good as reality. Whether the consumer compares the long waiting time on the Internet to long lines in the branches, or if wine tastes better sealed with a cork, or if a Rolls-Royce rides more smoothly if it has a leathery smell, it is important that a brand's sensory touch points be kept alive. These touch points should be maintained and enhanced since they are what gives the brand its unique blueprint.

Discarding valuable sensory touch points will downgrade your brand. Your primary objective therefore should be to ensure that all the historical links and associations connected to your brand are supported. If you fail to do this, you are at risk of losing some of the strongest competitive advantages of your brand.

Loud messages don't stand a better chance of being heard in today's world. However, a message supported by appealing to several senses stands a far better chance of breaking through. The *BRAND sense* study confirms that the more positive the synergy that's established between our senses, the stronger the connection made between sender and receiver. It's as simple as that!

A fabulous photograph of a freshly picked apple glistening with morning dew might attract you. But if you could smell it, and hear it being sliced, it might convince you. Add to the picture a texture that might even give you a taste, you would certainly be prepared to buy it.

Branding has always been about establishing emotional ties between the brand and the consumer. As in any relationship, emotions are based on information gathered from our senses. Online dating agencies are one of the Internet's most financially successful operations. You start with a photograph, you may move on to hearing a voice . . . if both convey a positive sense, you may be persuaded to actually talk to the photograph at the other end of the telephone line. All these cues may be pointing in the right direction, but unless there's a physical presence you will never know if you even want a second date. All our senses are required to fully evaluate our choices.

Brands are no different, although up until now communication has been confined primarily to the 2-D model with an occasional scratch-and-sniff scent added to the odd perfume ad. This is as advanced as it ever became. However, if you want consumers to access more dimensions of your brand, you need to move from a 2-D model to a 5-D model. The too-hard nut has to be cracked. Every sense opportunity should be explored. Don't hesitate to venture into the worlds of taste, touch, and smell, because the purpose of sensory branding is to ensure a systematic integration of the senses in your communication, your product, and your services. This will stimulate the imagination, enhance your product, and bond your consumers to your brand.

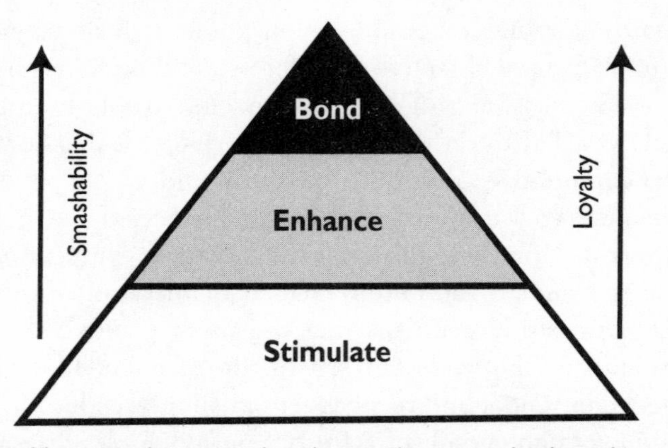


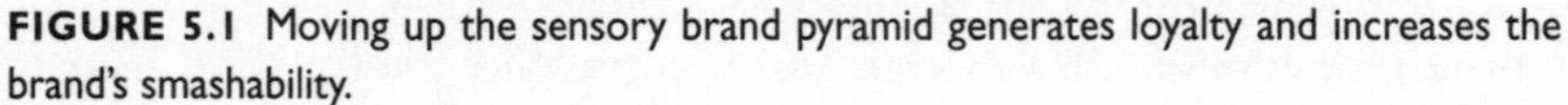
**FIGURE 5.1** Moving up the sensory brand pyramid generates loyalty and increases the brand's smashability.

## *Stimulate*

Imagine this. You're walking down a city street on a summer's day. The air is hot, the traffic's busy, the exhaust fumes are hovering, and there on the next block you see an ice-cream shop. As you get closer, a waft of fresh-baked cones greets you. Without giving it any conscious thought, you find yourself drawn into the store. Ice cream strikes you as the perfect antidote to the heat and the traffic. Before you know it, you're continuing your journey with a cool, delicious ice-cream cone in hand.

Sensory branding aims to stimulate your relationship with the brand. You could say it optimizes impulse purchasing behavior, sparks our interest, and allows emotional response to dominate our rational thinking.

Two sets of stimuli can take place—branded and nonbranded. That ice-cream shop aroma could have belonged to any number of brands. But a combination of aroma and the signage on the store helped create a brand association with a refreshing product. That association may arise unbidden the next time you're walking down the street in the heat. So what began as an unbranded experience is likely to convert into a branded one.

A branded stimulus not only motivates impulsive behavior, it also directly connects emotions to the brand. Take another hot day. You're sitting outside at a restaurant in front of a glass of ice with a piece of lemon, and hearing the sound of a drink whooshing open and being poured in the glass. Chances are you're thinking of Coke, because 78 percent of those surveyed have positive associations with the bubbling sound of an opening Coke can—or bottle. The fact is that this distinct sound of Coca-Cola has similar strong associations across the world. It's an association not dissimilar to Pavlov's dog, who got to expect his meal each time the bell rang.

In Scandinavia, a home-delivery ice-cream company has taken the Pavlov's dog metaphor to another level. A small blue van drives up and down neighborhood streets ringing a bell. After almost thirty years of riding around the towns, 50 percent of the population associates the sound with ice cream—not just any ice-cream, but more specifically with Hjem-Is ice cream.

To achieve a branded stimulus is one of the most difficult aspects of a sensory relationship. It's not intuitive, and it takes time to form. It requires constant reinforcement between need and specific brand. Branded stimuli create long-term loyalty. Nonbranded stimuli generate impulsive, yet nonbranded behavior patterns.

## *Enhance*

Just as a hologram allows you to see the same figure from different angles, so sensory branding allows you to see different dimensions of a single brand. Philippe Starck is a French designer who has turned his hand to everything from toilet-roll holders to hotel interiors, Puma shoes, and Microsoft's latest optical mouse. As diverse as this range is, each item redefines the traditional appearance with good design while maintaining function. His designs bring a fresh vision to everyday items, forcing us to see them in a whole new way.

Each sense that's utilized potentially adds a new dimension to the brand. The Bang & Olufsen BeoCom2 phone not only looks distinctive, but it has its own ring tone. The sound is considered more "human" than current electronic-based rings, and was created by Danish keyboard player and composer Kenneth Knudsen, thereby adding a surprising element to the highly branded Bang & Olufsen experience. Similarly, over the past few years Ford has increased its focus on sensory branded enhancement. They have used computers to generate a signature rumble for the engine of its redesigned F-150 truck.

As with stimuli, there are two levels on which enhancing brands operate—branded and nonbranded. Nonbranded enhancement is a common phenomenon, and is increasingly being seen (or smelled, felt, touched, and heard!) across the Fast Moving Consumer Goods (FMCG) category, which includes everything from candles to bed sheets. Toilet paper manufacturers are adding fresh-smelling scents to enhance impressions of sanitation and quality. This almost generic enhancement may affect the perceived quality of the product, but does little for the brand. The most effective strategy therefore is to secure a branded enhancement which, unlike a nonbranded enhancement, will reflect the brand, add distinction, and differentiate that brand from all the others on the shelf.

Texas Instruments has developed an exclusive touch for the keys on their calculators. This is a branded enhancement that makes the feel of a Texas Instruments calculator completely different from that of any other calculator. Similarly, Apple users are well acquainted with the Apple function key that replaces the generic "Ctrl" familiar to PC users. Loyal Apple users know that every shortcut involves the logo: for example, you press Apple + C to copy, and Apple + S to save. The name is part of the function and becomes part of the thought processes of any Apple user. The action is both branded and intuitive, making it a perfect example of branded enhancement.

## *Bond*

The ultimate goal in sensory branding is to create a strong, positive, and loyal bond between brand and consumer so the customer will turn to the brand repeatedly and barely notice competing products.

IBM ThinkPad notebooks have managed to create just such a bond. These laptop computers navigate with a TrackPoint mouse.

The system has been trademarked by IBM, ensuring their competitors do not duplicate it, and it keeps ThinkPad users loyal to the brand. Those who become accustomed to this system find it very difficult to switch to a touch-pad system of navigation.

Navigation is one of the most powerful ways for a brand to bond with the consumer. Whether it's an IBM TrackPoint mouse, a Nokia cell-phone menu, or the Apple's icons and setup, once a navigation system has been mastered there is a natural resistance to learn a whole new system. The process has become intuitive and most people are reluctant to interrupt their day-to-day flow.

## *Purpose of Sensory Branding*

Sensory branding will add four important dimensions to your brand:

1. Emotional engagement
2. An optimized match between perception and reality
3. Creation of a brand platform for product extensions
4. Trademark

### 1. Emotional Engagement

Sensory branding offers the potential to create the most binding form of engagement between brand and consumer seen to date. The goal is a very loyal relationship built up over a long period of time. In order to establish this bond, the sensory appeal has to have two essential ingredients: it must be unique to your brand, and it must become habitual. Not all sensory branding initiatives will necessarily be able to generate such high levels of loyalty, but loyalty will result if the brand maintains a distinct sensory appeal that is not imitated by any competing brands.

### 2. Optimize the Match Between Perception and Reality

Before Carlsberg released its new plastic bottle, it was tested. Danish focus groups were overtly aware of the change in the sound of the bottle opening. As a result of the findings, a special campaign was established to prepare customers for the change in sound and tactile feeling.

Too many brands allow too wide a gap between consumer perception and product reality. To narrow this gap, flower shops add artificial fresh-flower odor to the store. The genetic code of supermarket grapefruit has been tampered with to make the fruit easier to pack, and this has altered the taste. Consumers now expect their grapefruit juice to taste more like the supermarket variety. A juice company that grows its own grapefruit needs to be mindful of this "supermarket" flavor and get their juice to resemble it.

If quality is associated with substantial weight, then substantial weight must be added. If the rolling down of the automatic car window doesn't sound like "quality," the sound will have to be altered. In all cases, reality must shift closer to perception. The goal is for the reality to match, if not actually exceed, the consumers' perception.

### 3. Create a Brand Platform for Product Extensions

As each brand develops brand extensions, the links between the many products may erode unless a careful brand-extension strategy has been put in place. Consumers are able to make illogical leaps in product variety—for example, Caterpillar tractors and Caterpillar shoes. The products' link is established in avenues that extend beyond just the common logo. In the case of Caterpillar, the brand value is "masculinity." This has been translated into the use of materials—rubber, metal—colors, and positioning strategy. Sensory branding will create the emotional link between product extensions using the sensory touch points that are consistently repeated across each new product category.

### 4. Trademark

The challenge brands will face in this new century will be their capacity to protect their identity from competitors. Sensory branding will become the best way to do this. Almost every aspect of a brand's sensory appeal can be trademarked. Trademarkable components are known as trade dress. A trade dress is how a product smells, sounds, feels, tastes, and is shaped. Each component has to be distinct. The challenge is by no means easy. Harley-Davidson lost a court case trying guard their specific sound. In their case their sensory touch point belonged to a nonbranded type of engine, and so they could not claim the sound as exclusively their own.

## Sensory Translation

What makes one brand significantly more successful at a multisensory approach than another? In the majority of cases, successful brands have been able to implement a well-planned strategy executed over a long period of time—often decades.

We discussed a few of these brands in chapter 2: Singapore Airlines' Stefan Floridian Waters sprayed on hot towels and distributed to passengers, the distinct crunch of Kellogg's cornflakes, or the stylized Bang & Olufsen phone with its unique feel and sound.

Any company can build a sensory brand, one that engages in a multisensory dialogue with its customer. The first step is to optimize existing touch points, then add alternatives to the brand portfolio. And all the while, each aspect of this process needs to be continuously nurtured. Consistency needs to be maintained and authenticity vigilantly guarded.

# *Six Sensory Steps*

The *BRAND sense* study has examined hundreds of sensory branding cases from around the world. Despite the wide range of variation

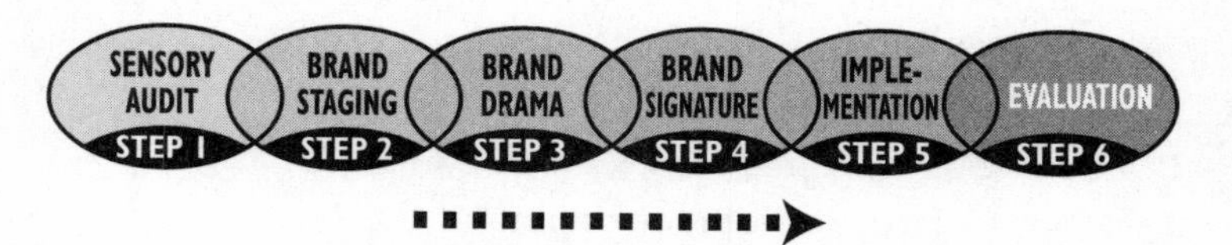


**FIGURE 5.2** Each step is interanked to ensure sensory synergy occurs during the development process.

in these cases, what's emerged in the analysis is that almost all of the most successful companies have followed a characteristic six-step process.

## Step 1: Sensory Audit

Evaluating your brand from a sensory point of view is far from easy. You've most likely known your brand for years, participated in dozens of focus groups, seen hundreds of reports on the brand performance, and worked through numerous prototypes before going to market. But how much time have you spent on the sensory components of your brand? How familiar are you with the sensory aspects? Do you believe that blindfolded consumers would recognize your brand? How strong is your idea of a sensory platform for your brand? What is the current status of your brand? To what extent can your brand leverage its sensory branding appeal? What needs to be done for the brand to achieve its optimal potential? Is your brand smashable?

A lot of questions, for sure, but they need to be asked before you begin because it is vital that you measure your brand's current ability to perform. You may very well be on the right sensory track without actually knowing it. Alternatively you might find yourself in a situation where the work currently being done on the brand will turn out to conflict with future plans to optimize the multisensory brand platform.

After conducting *BRAND sense* focus groups with hundreds of consumers across thirteen countries, I've established a set of ground rules for analyzing the current state of a brand. Additionally we have analyzed the world's top two hundred brands from a multisensory branding platform—their ability to handle five senses, as well as build and maintain a future sensory brand platform.

What your brand should be aiming for is sensory excellence. In order to achieve this, you need to carefully evaluate the following criteria:

1. Leveraging existing sensory touch points
2. Synergy across sensory touch points
3. Innovative sensory thinking ahead of competitors
4. Sensory consistency

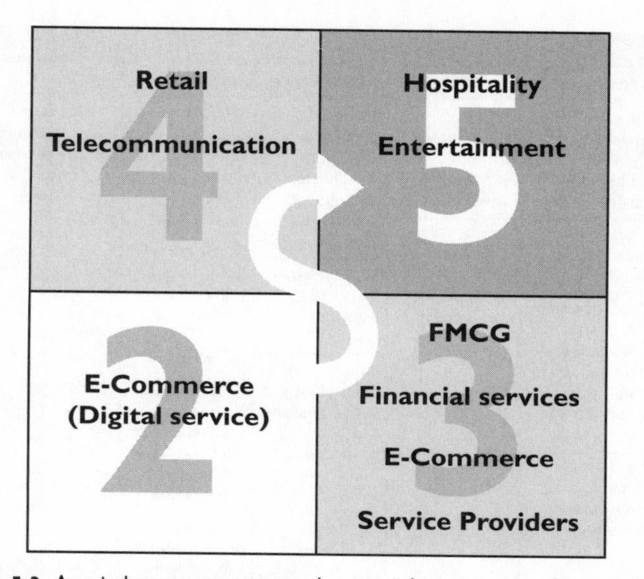


**FIGURE 5.3** Any industry category can leverage the minimum of two senses. Most brands should aspire to increase the number of their sensory touch points.

5. Sensory authenticity
6. Positive sensory ownership
7. Constant progress across sensory touch points

**1. LEVERAGING EXISTING SENSORY TOUCH POINTS** Because of the nature of their product or service, few brands are able to secure a total five-sensory appeal. It's clear, though, that any brand will appeal to at least two senses; and it's almost always possible to appeal to three.

Each brand should be audited according to our previous criteria: stimuli, ability to be enhanced, and bonding potential.

The objective of leveraging existing touch points is to optimize your branding platform. The more sensory bonding components in a brand, the stronger the foundation. If your product has limited sensory appeal, your goal should be to focus on establishing branded stimuli, branded enhancement, and bonding components. Should your brand already have several nonbranded stimuli, then your goal should be to convert these assets into branded components that

| STIMULI | | ENHANCEMENT | | BOND |
|---|---|---|---|---|
| Nonbranded | Branded | Nonbranded | Branded | |
| How does your brand currently fit into a multisensory model? | To what extent is your brand currently leveraging the senses? | Are you leveraging sensory branding for your current product development? | Signals from your brand need to become distinct and absolutely yours.<br>Or, for example, Heinz ketchup. It has a thick bottle—associated with the thickness of the sauce. | To what extent have consumers come to depend on your unique sensory touch points? |
| Does your brand have a smell? | Which components are totally associated with your brand? | How directly is it associated with your brand? | You may also need to thump it a few times to get the quality sauce to flow.<br>The more distinctions you can isolate as belonging entirely to your brand, the better. | |
| Would the store where your product is sold have a distinct aroma that would attract passing trade? | | Examine each touch point and see what can be uniquely applied to your brand. | | |
| Does your brand have a background soundtrack? | | | | |

**FIGURE 5.4** Questions that need to be asked during the stimuli, enhancement, and bonding process.

become your brand's own stimuli and enhancements that will foster more extensive bonding.

Far too many brands have many nonbranded sensory components. By branding them you can depend more on customer loyalty, rather than hoping the consumer will fortuitously stop at your product for her impulsive purchase. The object is to be aware of the dynamic of all your sensory touch points with a view to making them exclusively your own.

**2. SYNERGY ACROSS SENSORY TOUCH POINTS** Securing a synergy across sensory touch points is essential. It is evident from the *BRAND sense* study that a synergy between our senses does indeed take place, with some startling results. A sensory synergy can potentially double the effect of your brand communication.

**FIGURE 5.5** Sensory bonding is the ultimate level a brand can achieve. For example, Apple's navigation has cemented a strong bond between consumer and product.

Singapore Airlines leverages its Asian heritage by using a beautiful local woman as their exotic symbol. This visually desirable image is then enhanced by the Asian-style music that is heard on all Singapore Airlines commercials and played in Singapore Airlines lounges. The sensory optimization really kicks in when the same music is played in the cabin before takeoff, and is combined with the distinct exotic aroma that's been designed especially for them. Match this up with the makeup, uniforms, and appearance of the hostesses, and you have an all-round synergy between the senses. Each channel has been optimized, and then interlinked with one another, so that 2 + 2 equals more than 4! Each channel makes sense, each channel reflects the core values of the brand, and when combined the results are positively powerful.

The question you need to ask is just how strong are the links between each of your sensory touch points? How interdependent are they? If presented individually, would a consumer still be able to recognize your brand as distinctly yours? If not, look to the consistency of your core values and communication, then seek to instill those qualities into each component.

**3. INNOVATIVE SENSORY THINKING AHEAD OF COMPETITORS** Owing to the nature of their product, some industries are more predisposed to leveraging sensory opportunities than others. The perfume industry is more advanced in this respect because it has had to be. Perfume makers have had to find ways of letting consumers know exactly what they're selling in a world that communicates messages primarily through sight and sound. They've optimized their bottles, they've scented paper inserts in magazines, they spray whiffs of their scent in department stores, and there's always a tester to give you direct contact with the real thing. Perfumeries have been so successful that every industry that sells a product in a bottle looks to them for inspiration. Since it was impossible for the perfume industry to show their actual product, they were forced to optimize the usage of "alternative" channels, in a way similar to the blind or deaf person whose remaining senses are more strongly attuned than normal. The question is then what would you do assuming you were disadvantaged and could not demonstrate your product? How would this affect your communication, and more importantly, how could you involve your senses?

The automobile industry has managed to build a multisensory product over many years. A car is a completely controlled environment, giving the manufacturer a large amount of control over it. No detail has been overlooked, from how the control panel feels, to the shape of the car, to the fabric of the seating, to the sound of the doors and engine, with the new-car smell added as the icing on the cake. Just as the perfume industry has advanced considerably over the past decades by leveraging other senses to build their case, so the automobile industry has optimized and leveraged almost all our senses in every way.

Even though the perfume, automobile, and other industries have shown themselves stronger in leveraging the sensory appeal within the sphere of their natural merits, it doesn't preclude other less advanced industries from becoming excellent within their own industry category. Innovative sensory thinking ahead of competitors is relative to the business you belong to, so don't become discouraged when making a comparison with the more advanced industries.

**4. SENSORY CONSISTENCY** Consistency! Consistency! Consistency! This is the mantra for success. Consistency doesn't mean that things have to stay the same. On the contrary, it means that you stay true to your core values, which are expressed in your colors, your shape, your logo, and your sound. As in jazz, the melody keeps the thread while the music can take numerous turns and twirls. Consistency comes back to the very heart of your brand platform. If you've defined your consistent tactile, aroma, and visual expression, don't tamper with it—let it become mandatory in your brand platform to follow these guidelines. Let it become the law of your company. Make it part of your constitution that is impossible to alter unless major unforeseen circumstances force your hand.

Sensory consistency is what creates loyalty. It builds trust and generates repeat purchase, as people trust the familiar signals. Consistency generates history, history forms tradition and traditions leads to rituals. Chapter 7 will tell you much more!

**5. SENSORY AUTHENTICITY** Authenticity is another vital aspect of sensory branding. Things have to feel credible, real, and genuine—even if it's coming from an artificial place. That's the paradox. People don't necessarily want to step into The Matrix, where everything is an illusion.

Our *BRAND sense* focus groups show that the concern about corporate control of everything in our lives is particularly strong in the Northern European countries—Sweden, Denmark, and Norway. Southern Europe—Italy and France—are less fussed about it. Those in Thailand and the United States aren't terribly worried either. However, wherever this question was raised, the question of authenticity was important.

The fine line between authenticity and artifice is tenuous. An artificially created vibrating door that closes "with quality" is determined as authentic. We also feel comfortable about the artificially added "new-car smell." We accept the bird sounds coming out of the loudspeakers in Disney parks, and we may even buy a video of birds hopping about and chirping for our indoor cats. Perhaps we do so because these are components that fulfill the perfect illusion that we pay to become part of. Even though the sounds and smells may be artificial they are part of the real illusion.

Authenticity within a sensory branding context can be pinned to

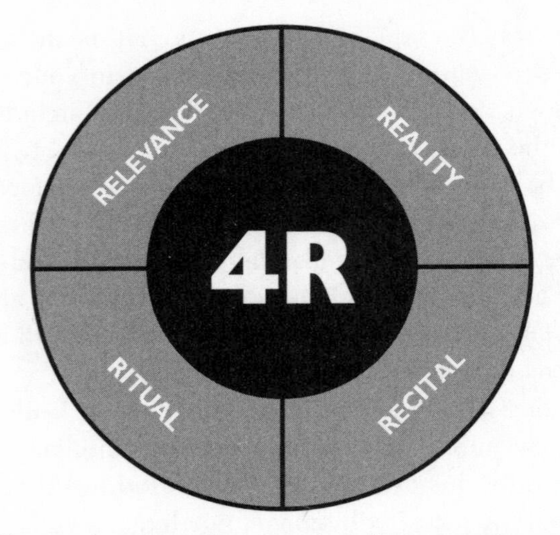


**FIGURE 5.6** Can your brand pass the sensory branding authenticity test?

*four* individual and very subjective factors. These four "R's" will help you identify the authenticity of your sensory approach. They offer a starting point, and serve as your authenticity guide.

Real + Relevant + Ritual + Recital

*Does It Feel REAL?* Plastic is such a versatile material that it can substitute for items as diverse as leather jackets and plumbers' copper pipes. The material came into its own in the mid-twentieth century, when it was used as a substitute for anything and everything. And so another expression entered the English language. People would say "It feels like plastic." By that they meant it didn't feel real. Language took it further, and described people perceived as insincere as "plastic."

Authenticity is a movable feast. A quick visit to the night markets in Bangkok or Beijing will put authenticity to the test. The fake Oroton wallets and Prada handbags generally feel real—some may feel plastic—but in all events, they're all "real" fakes. Apply a sensory solution to the vexing question. Does the item feel real? A fake wallet can still be an item of quality within the context of stall after stall of "plastic" wallets. Take the "Oroton" out of the night markets, and put it on the streets of Boston or Berlin, and within that new context its sense of authenticity becomes seriously diminished.

Real does not mean perfect. A Japanese billionaire who lives outside of Tokyo has the huge driveway leading up to his house swept every day. It takes a member of the household staff hours to do. And then when the job is done, and every leaf has been removed, he finishes the job by artfully placing a few leaves back on the roadway. This makes it look real. Things that are too perfect—too real—are perceived as being fake.

*Does It Feel RELEVANT?* Relevance depends on situation, context, and personality. Florist shops smell of flowers, and bakeries appropriately smell of bread. How about the smell of coffee in a bank? Perhaps coffee is a relevant addition to be served to waiting customers. How about vanilla-scented baby powder? Well, that also can be considered relevant because vanilla reminds one of innocence, and the subtle yet distinct smell seems relevant to baby smells.

Relevance is as flexible as the person who's perceiving it. However, there's a commonality in general perception, although major variants occur from culture to culture and generation to generation. A safety pin may be totally relevant to holding together a teenager's torn jeans, yet be considered completely inappropriate in the pants of a sixty-year-old. Context is everything in the relevance department. Therefore, results may vary according to the audience you are asking—or the audience your brand is to appeal to. Again, the context is essential for your evaluation.

*Has It Become a RITUAL?* Rituals have stories. And stories that contain rituals usually evolve over time. However, longevity doesn't necessarily define authenticity, although it goes a long way to support the authenticity of a brand. The Campbell Soup Company, which has been going since 1869, is not necessarily more authentic than Google, which has been around for less than a decade.

Rituals comprise formalized, usually repeatable and ceremonial actions. Even though they tend to be used in a religious context, there's every indication that commercial rituals will be increasingly prevalent in our social landscape. We pop champagne corks for a variety of celebrations. We celebrate Mother's Day with gifts and cards. We bring flowers to a grave. These are actions predominantly driven by ritual rather than by rational thinking. As such, brands would find it beneficial if they took their inspiration from this type of behavior.

Rituals are essential for strong brands. The more consumer-created rituals, the stronger reason for being, the stronger bonding between the brand and the consumer. Hard to imagine why canned laughter is so prevalent. But it's an integrated element in television, sitcoms and radio shows. Take away the canned laughter, and it would be conspicuous by its abence (but I think we'd be less comfortable with canned applause at a live show).

What rituals and routines exist for your product? Once there's ritual behaviors woven into the fabric of your product, you're right on track to establishing an authentic sensory approach.

*Is It Part of a RECITAL?* The most enduring fairytales are those that revolve around a story that conjures up emotion. These stories can fascinate and intrigue us, but they often engage elements that require us to suspend our disbelief. The witch's house in Hansel and Gretel or the wooden Pinocchio who comes to life could test most amateur scientists' fact books. But we don't care. Within the context we fully believe the story.

The context of the story lends credibility to the situation. How a brand was founded and how it's been used through time creates authenticity for the brand. If the sensory features of the brand are part of the brand's story, repeated recitals of this will give it a place in space, and thereby bestow it with authenticity.

*The Authenticity Score Card* The sum of your sensory branding activity and its ability to be real and relevant, as well as to contain

Visit DualBook.com/bs/ch5/authenticity to learn more about how to understand authenticity.

elements of ritual and recital, indicates the authenticity of your approach. Score each component between 1 and 5 (5 representing the highest score). You need a combined total of at least 10 to pass the authenticity test.

| AUTHENTICITY TEST | Car smell | Door sound | SIA |
|---|---|---|---|
| Does it feel real? | 5 | 2 | 2 |
| Does it feel relevant? | 3 | 5 | 3 |
| Does it represent a ritual? | 5 | 3 | 3 |
| Is it part of a recital? | 5 | 5 | 3 |
| **TOTAL** | **18** | **15** | **11** |

SIA= Singapore Airlines

**FIGURE 5.7** If you score more than 10 points, you can be fairly sure the sensory activity will be perceived as authentic by the consumer, and will therefore support the overall brand image. 1 = not at all; 5 = most definitely.

**6. POSITIVE SENSORY OWNERSHIP** U.S. citizens crossing the Mexican border are bringing back what they consider to be the "real thing": Coke in an old-fashioned green bottle. The fact that the Coke is produced in Mexico does not matter one iota. It's the bottle they're after. And they want the same bottle they drank their Coke in twenty-five years ago.

As we've seen, over time, brands have come to "own" certain sensory feelings. Disney owns two semicircular black ears, Nokia owns its tune, and Absolut the shape of its bottle. These are all positive components of the brand. In contrast, some brands own negative attributes—the smell of McDonald's oil, or the tinny sound of a Japanese car door closing. To own a part of a sense—whether it's a Tiffany blue or an Intel sound, a Colgate taste or a Singapore Airlines smell—is essential for a brand to become a truly holistic entity.

If you feel that your brand owns a distinct sensory touch point, you will be in a select group. Less than 1 percent of today's top brands can make this claim.

### 7. CONSTANT PROGRESS ACROSS SENSORY TOUCH POINTS

Tim Hortons, a popular coffee-shop chain in Canada, serves their snacks with real cutlery, Tupperware containers open with the same distinct whoosh that they did sixty years ago, but are these brands aware of the value of their service or product? Are they conscious that this has helped build a sensory loyalty?

Customers are the keenest defenders of a brand. Ask them questions. Let them help you identify your sensory assets:

- List all the sensory touch points (touch, taste, smell, sight, and sound) that you can think of when customers use this product or service.
- Identify the sensory components (touch, taste, smell, sight, and sound) which could adversely affect your brand if it failed to deliver according to the standards you have in place.
- Identify the most important sensory experience on your list, the one that is absolutely essential for your brand to fulfill.

**FIGURE 5.8** A truly successful sensory branding strategy represents these components.

Technology has enabled us to appeal to senses in a way that was not possible a few years ago. In the same way technology has also brought an extraordinary amount of clutter to the airwaves, forcing us to become more distinct than ever. Multisensory branding requires diligence and patience over a long period of time. The car industry has been developing sensory brands since the 1980s. The perfume industry has persisted too, sparing no expense in determining that the brand matches the perception of the consumer.

**8. SMASH YOUR BRAND** Last but not least, you need to determine if your brand is smashable—the process detailed in chapter 3. Remember? If I remove your logo, can I still recognize your brand? Just think about the music for Bond, James Bond. If United Artists—the production company behind the James Bond franchise—were to repeat the same title song in each movie, they would have failed. They would also have failed if the title songs were as different as Prince and Pavarotti. What they came up with is a type of tune with a certain sound, varying the instrumentation here and there as you would a twist to a cocktail.

Exactly the same can be done across any other industry category, from cell phone tunes to navigation icons. In short, every sensory ele-

| EVALUATION CRITERIA | Current status | Goals |
|---|---|---|
| Leveraging available sensory touch points | | |
| Synergy across sensory touch points | | |
| Innovative sensory thinking ahead of competitors | | |
| Sensory consistency | | |
| Sensory authenticity | | |
| Positive sensory ownership | | |
| Constant progress across sensory touch points | | |
| Smash Your Brand | | |

**FIGURE 5.9** By scoring the eight sensory critieria, a milestone can be established against which to measure any perceptible progress.

ment needs to be smashable, starting with the images . . . colors . . . shapes . . . name . . . language . . . icons . . . sound . . . navigation . . . behavior . . . service . . . tradition . . . rituals.

Playing with your brand's sensory platform is like tampering with its personality. It's serious business requiring a solid commitment from all stakeholders within the company. The goals you've outlined will help you set and maintain your focus.

## Step 2: Brand Staging

An old Scandinavian proverb says you will never become a leader by following another leader's tracks in the snow. Leveraging the concept of sensory branding might be one of the few ways for you to secure a truly competitive advantage. So, who do you learn from? In the world of sensory branding you may find it beneficial to look outside of your immediate industry. Rather than keeping an eye on your competitors, check out other industries involved in innovative approaches. They may prove to be the most valuable teachers.

Let's say your business is furniture. Looking to the global operation of Ikea would only really give you a benchmark to compare the success of your own business to. Rather redirect your focus to other industries with a different way of thinking. Likewise, Ikea might have more to learn from a Disney theme park than from a large furniture retailer. Disney could teach Ikea about holistic branding. From the moment the customer pulls into the parking lot until the very end when he's packing his purchases in the trunk, Disney could impart a few hints for Ikea on how to manage the flow.

Using Figure 5.9 as a guide, one of your objectives should be to identify companies that share similar challenges to your own, and companies that are leaders within their own fields that have leveraged the five senses. For example, Disney might be one of your sensory benchmarks even if you are a car manufacturer. You may look to Singapore Airlines as a benchmark for the second criterion, to achieve synergy across sensory touch points. Nike would be a good company to look to when you're wishing to examine constant progress across sensory touch points. Identifying a company as an example for each of the eight categories will give you a point of reference.

| EVALUATION CRITERIA | Current status | Goals | Sensory bench. |
|---|---|---|---|
| Leveraging available sensory touch points | | | Disney |
| Synergy across sensory touch points | | | Sing. Air |
| Innovative sensory thinking ahead of competitors | | | Apple |
| Sensory consistency | | | Chanel |
| Sensory authenticity | | | Rolex |
| Positive sensory ownership | | | Nokia |
| Constant progress across sensory touch points | | | Nike |
| Smash Your Brand | | | Coke |

**FIGURE 5.10** Sensory benchmarking is not about comparisons with competitors, but rather about identifying brands that have accomplished exceptional sensory progress within one of the eight sensory performance categories.

## Step 3: Brand Dramatization

See branding as a theater. Brand dramatization is all about the personality of your brand. Who are you? What feelings and emotions can be generated by enhancing your sensory appeal? What sensory priorities should be leveraged to perfect perception of the brand?

It is important that you identify to what extent each sense plays a role, then create a synergy between them. Perhaps your brand has focused exclusively on its visual attributes, but consumer testing shows that the tactile component of the brand is equally important. Take Kodak for example. More and more people are taking digital photographs. Yet despite this trend, consumers consistently report that they associate the tactile feelings of photographs with Kodak. Years of waiting for pictures to be developed, then experiencing the excitement of seeing them and reliving joyful occasions has left its mark. Photographic images and positive associations are still the province of Kodak, but will they be able to keep this sensory advantage when digital prevails?

Kodak has stayed in the game. They are the one of the world's leading manufacturers of digital cameras and have many tactile opportunities spread across their entire product range. Yet the consistency of touch from camera model to camera model, from color

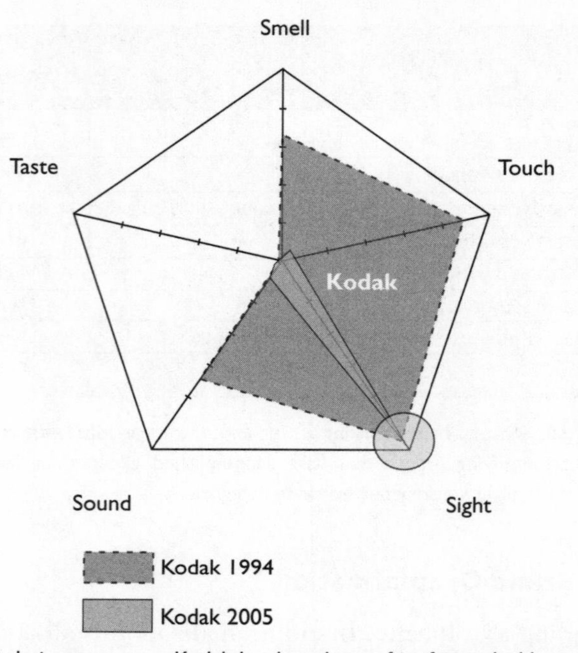


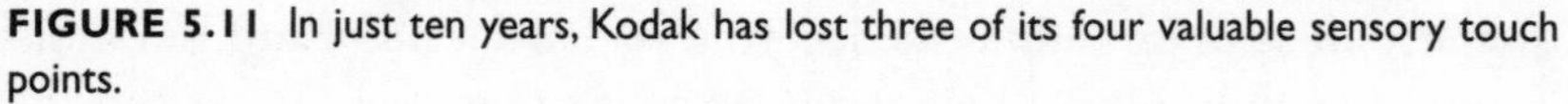

**FIGURE 5.11** In just ten years, Kodak has lost three of its four valuable sensory touch points.

printers to CD cases, is virtually nonexistent. Without identifying this valuable tactile feeling, Kodak is in danger of losing a major sensory branding opportunity.

If Kodak should become aware of their major sensory branding advantages, they may choose to change their focus, and upgrade the tactile features on their products, thereby ensuring many more years full of glorious Kodak moments.

The identification of the sensory touch points represented by your brand is essential for further sensory branding work. This helps to rebuild a new brand platform without destroying the invisible ties between your brand and the consumer.

**VALUES** The personality of a brand reflects its values. The more distinct the values, the better opportunity you have to create a distinct sensory appeal.

The goal is to convert each of your core values into a sensory touch point. Picking one or several senses, you should convert each into a concrete experience. As each sensory touch point is refined, so the emotional ties between product and consumer grow stronger. Here are some examples of how to optimize the sensory qualities of your brand by using the Stimulus, Enhance, Bond model.

---

**STIMULUS**

**Brand**: *Louis Vuitton* Visit any of the three hundred Louis Vuitton stores across the globe and you will find the identical display in the shop front window. Each month, on exactly the same day, the design is changed according to the global manual of window dressing. The LVMH group is placing all its emphasis on the visuals. Everything from the door handles to the wall texture to the packaging is carefully controlled. However, they've yet to begin leveraging the other senses. They might compare to Chanel's Rue Cambon store in Paris, where only the distinct aroma of Chanel No. 5 greets you. Nothing else is on display. You have to ask for what you wish to see!

**Values**: tradition, luxury
**Nonbranded stimuli**: classical music
**Branded stimuli**: specially composed Louis Vuitton signature music

**ENHANCE**

**Brand**: *Virgin* Despite the energetic values that Virgin undoubtedly portrays, they have only managed to achieve a limited synergy across the more than one hundred businesses under the Virgin umbrella. Virgin has, however, managed to establish some interesting touch points in many of the company's individual businesses. The release of Virgin Cola in 1996 reflects the perfect synergy between brand values and the ultimate sensory experience. Using the core values of the Virgin brand as a starting point, inspired by the shape of *Baywatch* babe Pamela Anderson, the curvy cola that emerged kept tipping over, forcing Virgin, as the joke goes, to refine their curves.

**Values**: innovation, fun, excitement, energy
**Nonbranded enhancement**: distinct, provocative shapes
**Branded enhancement**: recyclable soft drink bottle shaped like Pamela Anderson

**BOND**

**Brand**: *Colgate* For decades, Colgate has focused on its dental health product, which cleans. However, the brand has failed to ensure a taste-smell consistency across their product extensions. Ensuring a synergy between the brand's core values and the sensory appeal is essential to achieve synergy but also to establish authenticity.

**Values**: cleanliness
**Bond**: distinct smell and taste of toothpaste
**Bond extension**: distinct smell and taste used across all other products, for example, toothpicks, dental floss

---

## Step 4: Brand Signature

A brand signature is your unique statement. It is created when each sensory component is designed to form a complete Sensagram. Imagine you're a cell phone manufacturer. Let's look at a customer's sensory journey as she embarks on making a purchase.

Over time, the consumer has become of aware your brand. He's seen your commercials on television (audio and visual) and paged through a newspaper containing your print ads (visual). He may visit your website (audio and visual) to see what models are available. Next he stops in at a cell-phone store (audio, visual, tactile, aroma), where he gets to speak to an expert and try out the product.

Every consumer journey consists of a mixture of sensory touch points. Some may not be leveraged, others might be. Based on the criteria (values, bond, bond extension) set out to achieve sensory excellence, you must identify the most important steps that you need to take in order to convert the mixture into a distinct branded sensory experience.

In exactly the same way that people project their own individual personality, so should the brand. And in the same way that these personalities consist of a whole range of moods and attitudes, well, so should the brand. Again, in the same way that personalities engage in dialogue appropriate to time, place and circumstance, so should the brand.

But the goal is not only to leverage each and every touch point from a sensory point of view. Instead, like the baton passed in a relay race, one sensory touch point should lead to another. If the smell in the store is consistent with the smell when opening the box and later on when visiting the brand display at an exhibition, a sensory synergy takes place. If the sound used in the TV commercials and on the website continues in the store, greets you when you switch on the product or even when you contact the call center for help, a synergy takes place. However, this is only the first part of the sensory synergy story . . .

### SENSORY SYNERGY

What our studies show is that the same principles that are involved in traditional brand communications apply when building a multisensory brand platform. I call this cross-synergy theory. Cross-synergy theory says there has to be a relationship between a headline and an image to make any ad effective. If one of the two components is able to stand alone, cross-synergy does not exist.

Recently Airbus 340 ran an ad promoting their new more spacious seating plan. Instead of showing a picture of a seat, they showed a nut being cracked by a nut cracker. Beneath the image were the words "Middle seat anyone?" followed by the copy "Airbus 340. With no middle seat in business class." Neither text nor image would make any sense alone, but together their synergy formed an excellent ad. The same principle applies when leveraging our other senses.

From our study it is clear that if you show an image of a strawberry, then add a strawberry smell, the effect is good but not amazing. Yes, it creates attention, but it is not necessarily remembered because it is expected. It is when you combine the senses in nontraditional ways that true memorability occurs—like showing a picture of a car, and adding the smell of leather seats. Or showing an illustration of tennis balls and adding the smell of freshly cut grass. The links are not necessarily predictable, yet they are logical enough to justify the link.

## Step 5: Implementation

Once you have a fully prepared plan for your multisensory branding, you need to develop a step-by-step implementation proposal for every department affected by the strategy. Preparation of the proposal will necessarily involve the research and development department. In contrast with more classical marketing strategies and branding initiatives, sensory branding has to involve research and development, and sales and operations as well.

Once again it's important to stress that the sensory brand platform is as valuable an asset as the company's logo. Whether it be to develop a unique scent, a distinct taste, or a special tactile shape, the chances are you will need to seek external support in the sensory development process.

The implementation of your sensory branding strategy will consist of five distinct phases:

1. Development of sensory touch points
   - Generate a list of primary and secondary touch points, divided up according to branded/nonbranded stimuli, branded/nonbranded enhancement, and sensory bonding.
   - Benchmark each concept against a noncompeting brand that's been selected for innovation and excellence.
2. Testing concepts of sensory touch points
   - Benchmark and test the touch points in internal session.
3. Touch point integration
   - Incorporate the sensory touch points in the product specifications.
4. Testing with prototype
   - Test customers' perception of the sensory touch points.
   - Test authenticity of product.
   - Benchmark the results against competing and noncompeting brands.

5. Natural environment study
   - Ensure that the touch points in their real environment match those that have been created for clinical tests.
   - Check the status of the concept against the eight criteria of evaluation.

### Step 6: Evaluation

- To what degree does the revised sensory brand achieve the desired effect?
- To what extent is the sensory appeal loyal to its heritage?
- As a result of this sensory integration, is the brand still perceived as authentic? Already covered as part of the eight-point test earlier in this chapter.

Sensory branding is an ongoing process, and once in motion it requires constant monitoring. Measuring the process is essential and should be an integrated component in your program. In other words, checking the status of your brand from a sensory point of view should be just as essential as measuring your product awareness, loyalty, and market share.

## *Extensions and Alliances*

### Brand Extension

Brand extensions need to establish links that will help the consumer connect them to their mother brands. Consumers have reached new levels of sophistication, and product identity requires more than merely slapping a logo onto a piece of merchandise. People will not pay a premium price just because there's a familiar logo on an unproven product. A steady stream of brand extensions have made a wary public skeptical.

We're born to make connections. When a baby comes into the world, people stand around commenting on the resemblance between parents and infant. "It looks just like him . . . ," "She's got your nose . . ." It's in our nature. It doesn't stop at the cradle. We spend our life making these sorts of familial observations. We may even invent connections that don't exist; it's a way for us to connect the dots and make sense out of our reason for being.

We're always connecting dots. We check out family lineage. We follow threads of succession in successful businesses. We connect food first to region and then to nation. And the failure to make a relevant link is where most brand extensions fail. A logo alone is not enough. We have to be convinced that the very DNA of the brand we love is passed on to its progeny. Sensory branding might very well prove to be the link that binds the mother brand to her extensions.

In November 2003, the luxury cosmetic company Lancôme unveiled their first musical identity. The tune they chose to link their different products and services together was launched in Hong Kong at the Institut Lancôme's day spa. Jérôme Bartau, the director of development, stated, "The body's five senses are key but we have really only looked at scent, vision, and touch in our approach to consumers." Lancôme worked to find a sound which Bartau describes as "enveloping." This was so successful in the day spa that they used it on their website. It is their belief that the sound will strengthen Lancôme's brand values across their product portfolio.[1]

Another French company that's seen the value of sound and consciously set out to use it to reinforce their corporate identity is Air France. The airline recently underwent a massive refurbishing of their airline and their image. Part of this overhaul involved a conscious effort to sensory brand the airline. Elisabeth Ouillié, the brand manager for Air France, noted that, "The more a brand is present on all possible points of contact with a consumer, the stronger the brand is and the more the consumer retains a brand image."[2]

## Expanding the Sensory Universe

Exploring the sensory potential of your brand can provide major benefits. It's important that these explorations stay true to the core nature of your product. Quite frankly, in many cases the product itself will impose its own limits on this. For example, a television cooking show is simply unable to add smell to whatever it is they're showing on the screen. At least not yet! A bookstore will find it hard to appeal to taste—although many have overcome this obstacle by setting up cafés on their premises.

**BRAND ALLIANCES** Your extension opportunities will be defined by your product. However, you should not let this prevent you exploring every possible option to expand your sensory appeal without losing the focus of your core business. One way of doing this is to form a brand alliance.

Brand alliances—also called co-branding—happen when two brands team up to form a new venture or product. Alliances are formed between businesses with similar philosophies with an eye to attracting new customers and expanding their revenue base. Such alliances hopefully enhance the image of both brands.

There are two different types of brand alliances. In a functional alliance—also called ingredient branding—two items are necessary for a single product. The Diet Coke–Nutrasweet partnership or IBM–Intel are examples of this. The other type of alliance is symbolic. McDonald's serves Coke, or Qantas teams up with American Airlines to cover the globe. Across these two different alliance types, variations on a local, national, and global scale exist in order to gain maximum exposure and impact in the market.

A Sony-Kodak consumer brand alliances survey conducted by the

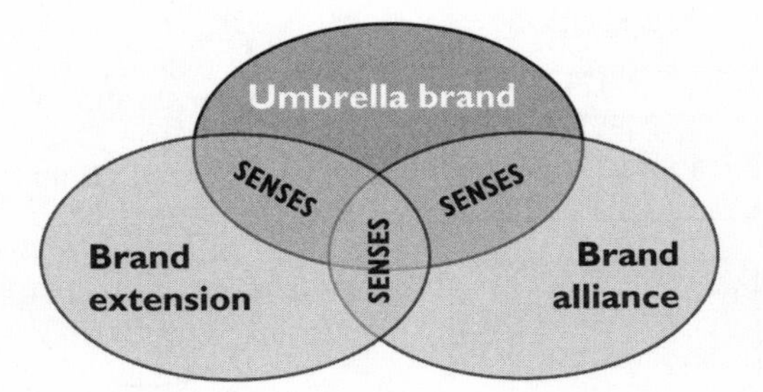


**FIGURE 5.12** Sensory links between the extended brand as well as the brand alliance partners must be synergetic. The creation of a sensagram for the umbrella brand will help form the best possible platform for further brand extensions.

American Marketing Association (1997) found that 80 percent of respondents would be likely to buy a digital-imaging product co-branded by Sony and Eastman Kodak. However, only 20 percent of those respondents were inclined to buy the product if it were branded by Kodak alone. Likewise, 20 percent said they would buy the product if it carried only the Sony brand.

If done correctly, brand alliances are effective. And this explains why they have become such a widespread phenomenon in recent years. In fact, over the past decade there has been a 400 percent growth in brand partnerships.[3] The top 500 global businesses have an average of sixty major brand alliances each. However, the rate of failure in brand-alliance partnerships is a large 70 percent.

Alliances require far more connection than combining logos. The links between the brands are often too weak, and in almost all cases sensory connections remain unexplored. In order for brand partnerships to be built on sturdy foundations, there must be a viable connection that creates a synergy from product to product. And consumers must be able to see the synergy from product to product. Multisensory

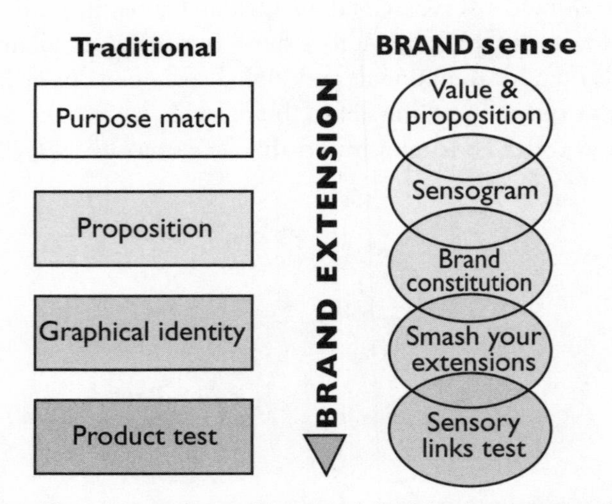


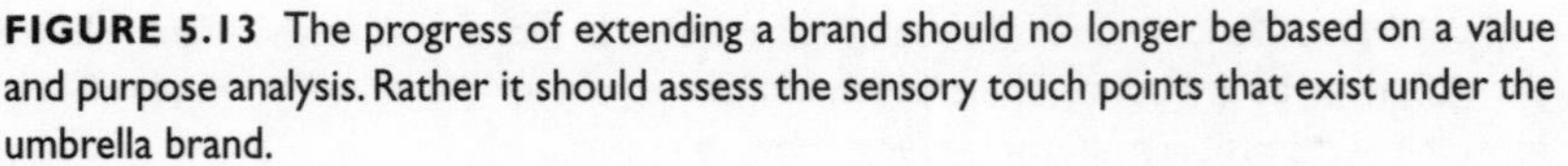
**FIGURE 5.13** The progress of extending a brand should no longer be based on a value and purpose analysis. Rather it should assess the sensory touch points that exist under the umbrella brand.

branding can provide the links necessary to form a sensory synergy which bonds brand with brand and alliance with consumer.

## *Case Study: Ferrari—Speeding Ahead Across the New Sensory Landscape*

As champagne corks popped at the 2004 Grand Prix World championship, another, slightly different Ferrari model was about to leave the factory. The recognizable features of the legendary racing car were in place—the brilliant red finish, the prancing black horse—symbols which over six decades have become synonymous with innovation, speed, and sophistication.

The new Ferrari 3000 had not neglected the sound either. The sound was there, and experts concurred that the 3000's speed paid homage to Ferrari's legendary status. To be sure, the Ferrari mystique was tangible on the day this new model was unveiled to the global press. There was an additional buzz, however. This new Ferrari had no wheels, and its engine was minute.

On this occasion, however, the symbols of quality and sophistication had been stylishly integrated into a sleek new computer. And it was this sensory appeal that guided the unlikely partnership of Ferrari and Acer, a computer manufacturer. Together they produced the Ferrari 3000, the world's first laptop fully clad in the patented Ferrari red. Although tenuous at first glance, the link between the two highly divergent brands is longstanding. Acer has been an official supplier of electronic components to the Ferrari racing team for quite some time.

It's certainly an innovative brand alliance, but more than that, it's an intriguing combination of sensory touch points. Besides the color, the notebook, like the car, has three coats of high-quality automotive paint and a brushed silver interior. The finish and the feel of the computer duplicate the finish and the feel of the car.

The same sensory synergy remains consistent in all Ferrari electronic products. In celebration of Olympus's sponsorship of the Ferrari Scuderia Formula One Racing Team, they released a Ferrari digital camera. Its body is Ferrari red, and it underwent five separate color checks in order to ensure accuracy. The aluminum parts are all hand-polished. The Ferrari 2000 camera reflects the fine lines of the Ferrari automobile. The high standards of both companies have been combined in a limited edition, high-quality camera that comes in a suede case with custom-designed strap.

The navigation for both the camera and the notebook is in sync with the design and navigation of the latest Ferrari car models. The notebook has leveraged an additional touch point—when you start it up, it sounds like a revving race-car engine. It's a sound so distinct that you can't fail to recognize it, and it's integrated into all Ferrari merchandise. You hear it when you log on to the Ferrari website, too.

The Ferrari 3000 is a fine example not only of how brands can team up to leverage each other's sensory strengths, but of how differentiation in one of the most competitive markets can be secured by optimizing sensory touch points previously neglected by the manufacturer.

By appealing to the tactile sense via the distinct Ferrari paint finish, and by leveraging the sound of speed and the branded color, Acer and Ferrari have managed to differentiate their notebook from the standard generic and basic grey models.

Interestingly, the product design and development process has not been characterized by a large R&D budget. Yes, there have been the standard technical upgrades, but by simply optimizing the existing touch points, Acer Ferrari has elevated its product from a pure performance-based notebook to a multisensory product that distinguishes itself from the pack on several platforms.

---

## Highlights

Project *BRAND sense* confirms that the more positive the synergy that's established between our senses, the stronger the connection made between sender and receiver. Discarding valuable sensory touch points will downgrade your brand. Your primary objective therefore should be to ensure that all the historical links and associations connected to your brand are supported. If you fail to do this, you are at risk of losing some of the strongest competitive advantages of your brand. You need to check for the following:

- ❖ The purpose of a sensory branding strategy is emotional engagement.
- ❖ Optimize the match between perception and reality.
- ❖ Create a brand platform for product extensions and ensure a trademark.

Companies that have successfully implemented sensory branding have generally followed a six-step process:

### 1. Sensory audit

Evaluating your brand from a sensory point of view is not an easy task. Leverage existing sensory touch points. There must be sensory consistency and sensory authenticity. Progress across sensory touch points must be constantly monitored, as well as securing a positive sensory ownership.

### 2. Brand staging

Securing a synergy across sensory touch points is essential. It is evident from the *BRAND sense* study that a synergy between our senses does indeed take place with some startling results. A sensory synergy can potentially double the effect of your brand communication. In short, each channel has to be optimized and then interlinked with one another, so that 2 + 2 equals more than 4!

### 3. Brand dramatization

Brand dramatization is about your brand's personality. Who are you? What feelings and emotions can be generated by enhancing your sensory appeal? What sensory priorities should be leveraged to perfect perception of the brand?

### 4. Brand signature

A brand signature is your unique statement. It is created when each sensory component is designed to form a complete Sensagram. In exactly the same way that people project individual personalities, so should the brand. And in the same way that these personalities consist of a whole range of moods and attitudes, so should the brand. Again, in the same way that personalities engage in dialogue appropriate to time, place and circumstance, so should the brand.

**5. Brand implementation**

Once you have a fully prepared plan for your multisensory branding, you need to develop a step-by-step implementation report for every department affected by the strategy. In contrast with more classical marketing strategies and branding initiatives, sensory branding has to involve research and development, sales, and operations as well.

**6. Brand evaluation**

This step involves stepping back and critically evaluating the new, improved brand.

- To what degree does the revised sensory brand manage to achieve the desired effect?
- To what extent is the sensory appeal loyal to its heritage?
- As a result of this sensory integration, is the brand still perceived as authentic?

This phase includes a step-by-step evaluation enabling the reader to measure the sensory performance of their brand.

Brand alliances are effective, which is why they have become so widespread in recent years. However, they're also risky. Multisensory branding can provide the links necessary to form a sensory synergy that bonds brand with brand and alliance with consumer.

---

## Action Points

- Identify how well your brand performs within each of the three categories: stimulate, enhance, and bond. Remember to separate your answers into "branded" and "nonbranded" categories. You should add the evaluation to the chart you created in chapter 2.
- Identify and evaluate the purpose of the sensory branding that you have in mind. Compare this to the evaluations

you've made in the previous chapters. Evaluate on four levels: emotional engagement, perception and reality, brand platform for product extensions, possibility of trademarks.

- Using the criteria outlined above, begin the six-step process.
- Structure these findings under eight headings:

  1. Leveraging existing sensory touch points
  2. Synergy across sensory touch points
  3. Innovative sensory thinking ahead of competitors
  4. Sensory consistency
  5. Sensory authenticity
  6. Positive sensory ownership
  7. Constant progress across sensory touch points
  8. Positive sensory ownerships

- So far you've only analyzed your competitors—it's now time to identify the companies you admire. Which in particular are strong within areas where your brand currently has showed to be weak according to your sensory branding assessment? These companies will serve as your future benchmark.
- What does your brand currently stand for according to your customers? Do you in fact represent some sensory advantages which you haven't been aware of? Combine the results from your focus groups with the core values that you want your brand to stand for.
- Identify each of the touch points your customers have with your brand when they purchase and consume your products. Identify the type of signals you intend to expose the consumer to during each step as well as the synergy to be generated across each touch point.
- Multisensory touch points can offer the link between two brands joining to form a partnership.

CHAPTER 6

# Measuring Senses

**FACED WITH NEW PRODUCT FAILURE RATES** around 80 percent, many of the world's largest brand marketers are realizing that it is their big, successful, established brands that represent their best opportunity for profitable growth. Compared to new products, established brands typically offer the advantages of an established customer base, less volatile revenues, and good margins. Success, however, is never guaranteed. In order to ensure continued growth marketers must find new ways to differentiate their brands from competitors. In the preceding chapters we have highlighted how the senses can be used to provide many brands with a more compelling brand offering that will help increase consumer loyalty and ultimately ensure profitable growth. In this chapter we examine the results of sensory audits conducted for six different brands, ranging from colas to home entertainment systems and soap. The *BRAND sense* research highlights each brand's existing sensory profile and helps us examine how these and other brands tune their sensory profiles to evoke emotions that best fit the brand's positioning.

## *Why Do I Love You?*

The objective of the *BRAND sense* research was to create an inventory of the sensory impressions for a number of different brands, highlighting how strongly the impressions came to mind, whether they were positive or negative, distinctive or not, what emotions they evoked, and how they affect existing users' loyalty to the brand. This was not an easy task. Have you ever asked someone why they love you? Were you disappointed with the answer because they just said "Err . . . because I do"? It is really difficult for us to break down any emotionally based experience into words, whether it is to describe why we love someone or why we bond with a brand, and yet that is just what we need to do in order to conduct a sensory audit. The problem is exacerbated because people tend to think about the sensory experience of brands only in terms of the primary sense involved in the experience: food tastes good, sound systems sound good, athletic shoes feel good. In order to create a sensory inventory for a brand we needed to get beyond the primary sense and tease apart the sensory experience to identify its component parts.

Faced with objectives like these, market research practitioners typically resort to qualitative research techniques. Working with small groups or individuals, a trained moderator will use different techniques to tease apart the brand experience to get a good understanding of each brand's touch points, sensory impressions, and the reaction to those impressions. That is what we did during the first stage of the *BRAND sense* research, but while this type of approach can provide great depth of understanding, it does not quantify the number or scale of the sensory impressions or the degree to which they affect brand loyalty. Think of it like using a recipe. Qualitative research identifies what ingredients are used to create the dish, but you still need to know how much of each ingredient is required to produce the optimum result. For that we need to use quantitative techniques, in this case survey research.

Most of you will be familiar with survey research. Everyone has been stopped when shopping or phoned during dinner by someone wanting to ask a few questions. The problem is that unless the questions are put in some form of context, they can often sound completely inane. If we came straight out and asked people "What does Mercedes taste of?," for instance, it is highly unlikely that we would get any sensible answers, and people would not even finish the survey. What we needed to do was create a survey design that would start by getting people to think about their own senses, and which ones they were most responsive to, then lead on to talk about differences between brands based on the senses, before finally getting into the meat of the questionnaire by focusing on the senses for specific brands. For each brand we first identified how strongly each sense came to mind. Then, for each sense, we asked:

- Did the impression make them feel positive or negative about the brand?
- How distinctive was the impression?
- What specific memories and emotions were related to it?

The responses to these questions will never match the insights provided by qualitative projective techniques. In qualitative research we can get people to ascribe tastes to Mercedes because they have been introduced to the topic in-depth and realize that they are being asked to draw parallels, not literally taste the car. Our survey questions do, however, allow us to create an inventory for each brand of existing sensory impressions. The next step is to identify how these impressions relate to brand loyalty.

## *Measuring Loyalty*

In our survey we asked frequent brand users how likely they were to consider the brand next time they buy a soft drink, visit a fast food restaurant, buy a home entertainment system, and so on. The top point on the scale was "It's your first choice." This scale has been used to accurately predict purchasing behavior in numerous product and service categories around the world.[1] The more people select

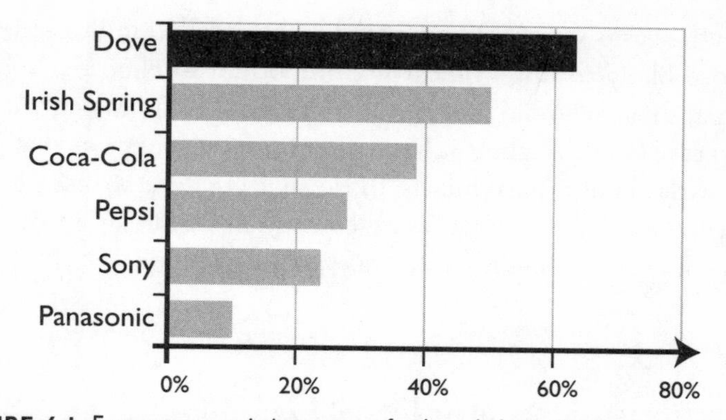


**FIGURE 6.1** Even among existing users of a brand the willingness to agree that the brand is their first choice differs widely. Sensory impressions help determine how willing people are to make the brand their first choice.

"It's your first choice" the more sales revenue the brand will generate (provided it has got its business basics right).

As can be seen from Figure 6.1, even though all the people surveyed were regular users of the different brands, there were big differences in the proportion who said the brand was their first choice. This data was based on all surveys conducted in the United States, UK, and Japan, and in all cases the differences between the brand pairs were statistically significant.

Clearly categories differ in the degree to which people are more concerned with choosing a specific brand than getting the best price. In the United States, people are more concerned with choosing a specific brand of soft drink than a home entertainment system (68 percent versus 41 percent). This difference is driven by the absolute price and the degree to which people believe that brands truly differ. The price point is low for soft drinks and people believe that they do differ, so loyalty tends to be high. The *BRAND sense* survey results shown in Figure 6.2 confirm that the more senses that come to mind, the more likely people are to make their existing brand their first choice. The conclusion is based on U.S. and UK data only, since people in Japan tend to be more sensitized to the senses and mention them more than their Western counterparts, clouding the comparison.

In all the cases we examined, people who recalled multiple senses were more likely to think that it was important to choose a specific brand than those who did not. But is this a chicken or an egg? Academic literature would suggest it is an egg.[2] Research suggests that prior sensory experiences, particularly those that create a strong positive emotional response, do affect brand evaluations. But can we find any evidence in our data that it works this way in real life?

## *Analyze This*

Do remembered sensory experiences really help create differences in brand loyalty? Further, can secondary senses, which are not directly related to the consumption experience, help create synergies that contribute to brand success? To attempt to answer these questions, we have used a statistical analysis to highlight which sensory associations appear to affect loyalty for an individual and how. To do so, we need to allow for the fact that the senses will likely work through one of the drivers of brand success: great brand experience, clarity, or leadership.[3] The drivers shown here represent the degree to which people agreed with a number of different attitudinal statements used in the survey research.

**Great experience**: you really enjoy using/drinking X more than other brands, it appeals to you more than other brands, or it is the highest quality brand.

**Clarity**: X has a very distinctive identity or is different from other brands.

**Leadership**: X is setting the trends, is the most authentic, or most popular.

The statistical approach we used is called Sequential Equation Modeling (SEM). Like any regression technique, it allows us to identify the degree to which an explanatory variable, like how much people say they enjoy drinking Coca-Cola, relates to a dependent variable, like how strongly they would consider the brand next time they buy a soft drink. What distinguishes SEM from other approaches, however, is the ability to test for sequential relationships or "paths" to consideration. SEM allows us to test whether or not taste leads to a perception of enjoyment which in turn affects consideration. For each path the model returns a coefficient indicating how strong the influence is on the variable to which the path leads. Paths that are insignificant are not reported. The model indicates how important each variable is, not how many people agreed that each brand possessed that property, which we will look at separately below.

As you might expect, each brand and category can have a complex and very different set of relationships. In order to make comprehensible comparisons between brands, we have developed a common template to describe how the senses might influence loyalty. This dramatically risks simplifying the true situation for each individual brand, but makes it far easier to document here. This general path model has been applied across several product categories, including colas, home entertainment, and soap, covering nine brands in total. In each case the model has proved to be statistically robust and confirms that the senses do have a role to play in creating brand loyalty.

Figure 6.2 is a composite model based on all nine brands. As a result, the strength of the individual relationships is very dependent

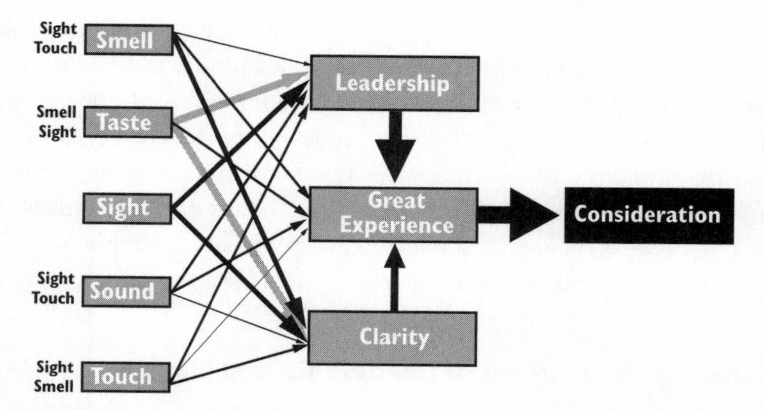


**FIGURE 6.2** Our analysis reveals how the different senses influence brand loyalty. All the senses, with the exception of sight, impact the three drivers of loyalty, which in turn affects consideration. Sight plays an important supporting role to the other senses.

on the brands and categories included in the research. For instance, due to the presence of food and drink brands in the set, taste has a stronger influence than the other senses. While all five senses have a significant influence on consideration in this overall model, rarely did more than three of them have a significant role in any specific brand model. Importantly, the results confirmed that in order to have the maximum effect on loyalty, the sense must come readily to mind and be positive and distinctive. So it is not simply a positive sensory impression that drives consideration. These results suggest that the sensory experience must create some form of differentiation, contributing to perceptions related to either clarity or leadership, in order to create real loyalty.

As the model shows, each sense can affect loyalty by one of three paths: leadership, experience, or clarity.

## Directly Impacting Perceptions of Experience

Given that we were conducting the research among frequent users of the brand, it is to be expected that great experience is the focal variable that affects consideration in every model. The degree to which it does so is relatively consistent between the two brands in each category, indicating that it is *how* the experience is created that is important to competitive advantage.

It is fascinating to note that sight does not have a significant relationship with brand experience across the brands, but has its influence through creating perceptions of leadership and clarity. Sight also plays a strong supporting role to the other primary senses. This may be a function of the categories involved, since only in the home entertainment category is sight a primary means of consumption, but it does suggest that sight is a less powerful influence on loyalty than the other more intimate senses.

## Creating Perceptions of Leadership and Clarity

Earlier sections of the book highlighted the role of authenticity, consistency, and clarity. Here we see the proof that these properties do matter. In all models leadership and clarity both affected perceptions of experience and were in turn affected by at least one sense. In all but two models leadership proved to have more influence than clarity, as shown here.

## *The Relative Impact of the Senses*

The different influences of the senses can be summed up to create an overall Loyalty Impact Score as shown in Table 6.1. The table indicates the degree to which each sense (with a significant role in the model) impacts brand consideration across the nine brands. The maximum score would be 1, indicating that there was a direct 1-for-1 relationship between the sense and consideration.

TABLE 6.1
**Loyalty Impact Score**

| | AVERAGE | MAXIMUM |
|---|---|---|
| Taste | 0.19 | 0.44 |
| Smell | 0.13 | 0.19 |
| Sound | 0.10 | 0.15 |
| Touch | 0.08 | 0.10 |
| Sight | 0.07 | 0.14 |

What this model does not capture is the influence of all the other marketing variables that come into play to determine loyalty: availability, packaging, warranties, price, promotion, and so on. This is sometimes apparent in the model results. Burger King users, for instance, are less loyal than McDonald's users, but this does not appear to be a function of the sensory experience. Rather it reflects the overall marketing and distribution power of the McDonald's brand.

### The Taste of Success

Even though there was relatively little difference in loyalty between Coke and Pepsi drinkers, our *BRAND sense* research suggests that the senses do have a role in creating competitive advantage in the cola category (Table 6.2).

**TABLE 6.2**

**Loyalty Impact Scores for Coke and Pepsi**

| | COKE | PEPSI |
|---|---|---|
| Taste | 0.44 | 0.43 |
| Smell | 0.16 | 0.15 |
| Sight | 0.08 | 0.05 |

Taste is the dominant sensation for both brands, with smell providing important cues to taste for some people. Sight plays an important supporting role. While the path model does indicate that the senses have an influence on loyalty, there is relatively little difference in the way they do so, or the degree to which the senses come to mind between the two brands.

The brands do differ, however, in the way people describe the taste sensation. For Coke the primary taste sensation is what some people refer to as "the burn" or "the bite."

> "The bite [Coke's taste] has compared to other colas. There is a slight sharpness between the beginning and the aftertaste that is rather distinctive."

> "Coke has a good blend of sweetness and sharpness. When I say sharpness I mean that it is not syrupy. It's got a clean taste to it and is very unique to Coke."

For Pepsi the primary taste sensation is less aggressive:

> "Light sweetness, smooth, no bite or strong aftertaste, tingling of bubbles pushes the flavor forward."

> "Light crisp taste and very carbonated. I hear it fizz in the open can. Also, very smooth taste."

Both sets of drinkers believe that their brand is equally distinctive, but slightly more Coke drinkers agreed that they felt very positive about the taste of Coke than Pepsi drinkers did of Pepsi. Perhaps the more challenging taste experience leads to a stronger emotional response, but the end result is that Coke has a slight advantage in

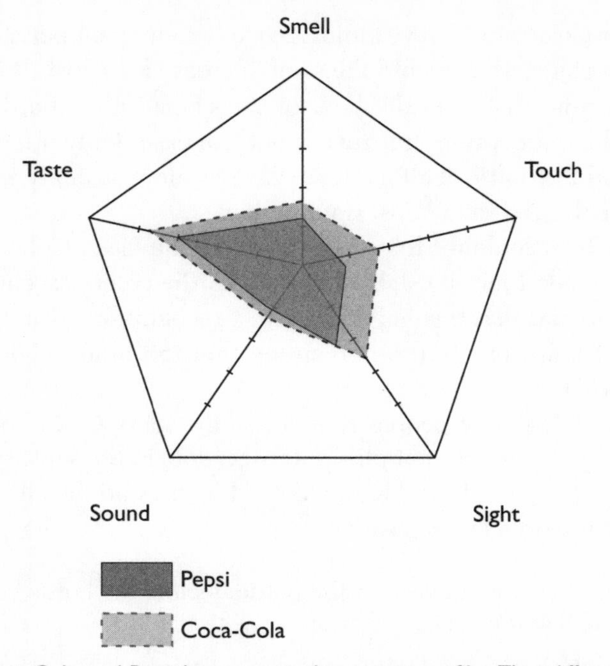


**FIGURE 6.3** Coke and Pepsi have very similar sensory profiles. They differ more in the nature of the sensory impression than how much they come to mind.

terms of people thinking the brand has a positive and distinctive taste. This result is consistent across the three countries included in our research (see Table 6.3).

TABLE 6.3
**Coke's Perceived Advantage**

| DIFFERENCE BETWEEN COKE AND PEPSI | |
|---|---|
| U.S. | +7 percent |
| UK | +10 percent |
| Japan | +9 percent |

One of the differences between the path models for the two brands was that sight had a significant effect on clarity for Coke, but not Pepsi. In turn, clarity had a stronger influence on the consumption experience for Coke than Pepsi. An increase in perceptions of clarity for Coke would have almost twice the effect on perceptions of great experience than would the same increase for Pepsi. It is tempting to assume that it is the look of the classic glass bottle that is driving this perception, but that is not the case. Only a few people mentioned the bottle at all. Instead, Coke is more strongly associated with the color red than Pepsi is with blue.

There is little doubt that over the years the Coca-Cola company has done a good job of establishing red as the color associated with their brand and that this helps provide a recognition advantage over Pepsi that helps create the perception that the brand is distinctive and different.

The fact that few people remember the glass Coke bottle may explain why touch does not play a stronger role in the sensory profile. Only a few people referred to the feel of the glass bottle, although the associations were always positive.

> **Touch**: "Touching the can or the bottle, feeling the bottle curves, knowing that it must be Coke."

> **Sight** (referring to Coca-Cola): "A Halloween party I had when I was thirteen and we got a case of the glass bottles as drinks. It was a very happy experience."

Today, the color of the can may differ between the two brands, but they feel just the same. In all three countries, the majority of Coke drinkers agreed that they much prefer the feel of the traditional Coke glass bottle compared to the Coke can. The same question asked of Pepsi drinkers evokes a less definitive response, highlighting the degree to which the Coke glass bottle is indeed a positive and differentiating property. Our research would, therefore, suggest that the glass Coke bottle is a lost sensory property that not only signaled authenticity but elicited positive memories for many people. What better way of triggering impressions of "the real thing" than using a sensory stimulus that creates an impression of reassuring authenticity by tapping into peoples' childhood memories?

## *See Me, Hear Me . . . Touch Me?*

Home entertainment systems seem like they should offer plenty of scope to accentuate the primary functions of sight and sound using the look of the design or touch. However, people were less likely to mention sensory impressions that were positive and distinctive than in other categories, and it is here that we find one of the weaker relationships between the senses and brand loyalty (see Table 6.4).

**TABLE 6.4**
**Loyalty Impact Scores for Sony and Panasonic**

| | SONY | PANASONIC |
|---|---|---|
| Sound | 0.15 | 0.13 |
| Sight | 0.11 | 0.10 |
| Touch | 0.06 | 0.05 |

While the nature of the category must play a large part in accounting for the lack of positive and distinctive impressions, one is tempted to ask if a single-minded focus on the picture and sound quality in the marketing has resulted in a very similar profile for both the brands included in our research. It makes one wonder if the focus on technical quality is just too rational for people to be involved with, or for that matter whether people can actually distinguish between brands on this basis. In all likelihood the technical quality of the picture has now crossed the boundary beyond which subtle differences in quality are simply not appreciated. Touch has a much weaker relationship with consideration, but may offer a far greater opportunity for differentiation that simply has not been leveraged.

Sony owners were more likely to think that sight, sound, and touch come to mind when they think about their brand than Panasonic owners did when they thought about theirs. The way that both groups described the associations, however, was undifferentiated. Pictures were clear and bright, sounds were crisp, rich, and clear, and touch smooth. So Sony's loyalty advantage owes little to actual remembered sensory differentiation and more to the fact that a higher proportion of people could bring some sort of impression to mind.

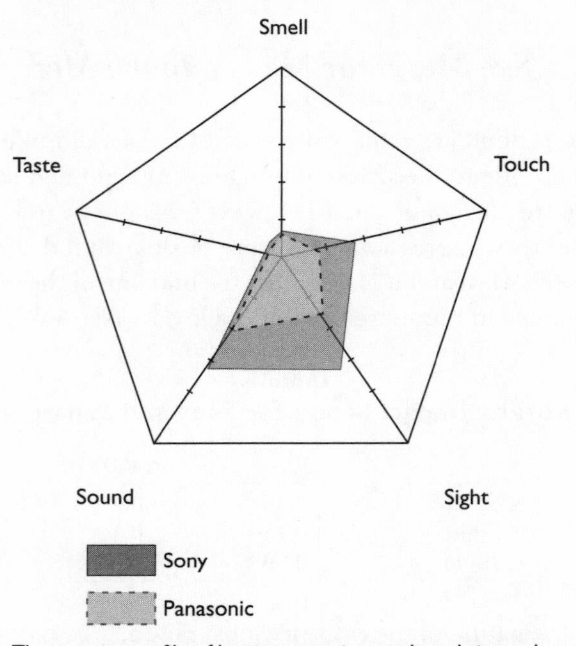


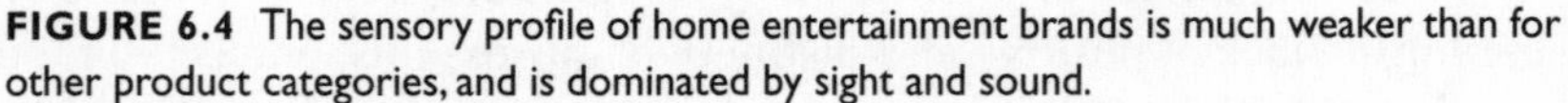

**FIGURE 6.4** The sensory profile of home entertainment brands is much weaker than for other product categories, and is dominated by sight and sound.

A few people talked about the style or design of the two brands, referring to them as looking cool, sleek, or simple, but compared to the richness of feedback from the other product categories we studied, one gets the impression that people were struggling to find anything to say. When asked directly, Sony users were more likely to agree that their brand had exceptional clarity of sound and images, and that it was distinctive and pleasurable to look at. The results, however, suggest that the design aspect of these brands is just not a salient feature to them.

Very few people commented on the design or feel of the remote, and those that did so were thinking in terms of ergonomics, like finding the right buttons in the dark, rather than the actual visual or tactile impression. Given that the remote is now the main means by which people interact with their system, surely it would pay dividends to ensure that it felt good to use and was not just some button-studded slab of plastic? When asked directly whether the buttons on their entertainment system's control panel and remote had a special, precise feel, few people said that they definitely agreed.

Overall, Sony is clearly the stronger of the two brands. This is reflected in higher consideration for repeat purchase among its current owners. The consequence for Panasonic is that the brand is less able to charge a price premium, reflected in the fact that its owners were more likely to agree that the brand was an acceptable price. WPP's BRANDZ research suggests that while one in four people are bonded to Sony in the United States, Britain, and Japan, far fewer are bonded to Panasonic, even in Japan. The question is, could design and tactile qualities play more of a role in creating perceptions of clarity or leadership and so increase brand loyalty for these brands in the longer term? I believe the answer is yes, since an examination of the two brands in stores suggests there are many potential differences that could be made more salient.

Sony has set up flagship stores in major cities. This has given them a wonderful arena in which to highlight the visual and tactile appeal of their products. Visitors are free to explore the line-up and handle each item. What an excellent opportunity to emphasize design and touch, to create competitive advantage! In order to reach a mass market, however, they might want to consider including a clearly stated focus on design in their advertising, in addition to sound and picture clarity. While most print ads do show the system being advertised, how many actually focus attention on what it looks like or feels like rather than on technical specifications and price? While the sense of touch is not an aspect of the brand that people will necessarily say is important to their final purchase decision, it provides additional cues to differentiate the brand and infer a greater degree of leadership.

## *Sensory Synergy*

The humble bar of soap has evolved over time to offer an amazing, multisensory experience! Many of today's soaps are handmade and include exotic ingredients, even fruit and herbs. But what of the mass-market brands that cannot indulge in such extravagance? Can they really establish a distinct and compelling sensory experience?

The answer, simply put, is yes, they can. In the U.S. survey we included the Dove and Irish Spring soap brands. Both have similar market shares, but are functionally different. Dove is positioned as a moisturizing beauty bar and might best be summed up by the word "softness." Irish Spring is a deodorant soap and can best be described as "invigorating." As we shall see, the two brands offer very different sensory experiences. Together the two brands highlight the ability to use the senses synergistically to create total clarity, supporting the overall positioning of the brand (see Table 6.5).

**TABLE 6.5**

**Loyalty Impact Scores for Dove and Irish Spring**

| | DOVE | IRISH SPRING |
|---|---|---|
| Smell | 0.11 | 0.17 |
| Touch | 0.10 | 0.06 |
| Sight | 0.08 | 0.07 |

Smell, not touch, has the most influence on consideration in this category, and, as we shall see, this is a more important driver for Irish Spring than for Dove. While both brands are equally associated with distinctive and positive smells and sights, these have been finely tuned to create experiences that are as far apart as possible.

Ask an Irish Spring user what smell most comes to mind about the brand and she will typically say something about freshness:

> "Fresh, outside spring air, light, airy."
>
> "It has a clean meadow smell to it."

Dove, on the other hand, offers a more subtle smell:

> "Dove has a very unique clean and fresh smell to it, with a subtle, not overwhelming, odor."
>
> "Dove is a pure soap and I smell its cleanness and purity when using it."

There is no doubt that Irish Spring has the stronger, more intense fragrance. Not only were its users more likely to think that smell

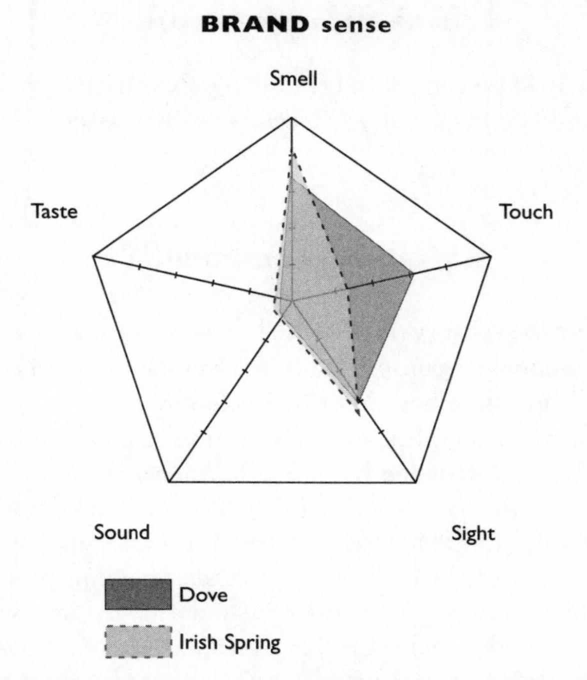


**FIGURE 6.5** Irish Spring and Dove use the senses to create very different brand experiences that are totally consistent with each brand's positioning.

came very strongly to mind compared to Dove users, they were more likely to refer to it as "an aroma bath" and talk about "the smell lingering after a shower." When we asked people in general which sense was most important in choosing between soap brands, 71 percent picked smell, which helps to explain why smell plays a stronger role in making Irish Spring users loyal to that brand.

It is in respect to the association with touch that Dove really stands out from Irish Spring. Far more Dove users think that their brand's touch is both more positive and distinctive. As a result, Dove users are firmly convinced that the brand feels soft and creamy (91 percent agree compared to only 56 percent for Irish Spring users asked about their brand).

With regard to sight, the two brands again offer different sensory experiences, but ones which augment and add clarity to the overall experience. Irish Spring's green and white striations fit with the connotations of freshness. Dove's white, curved oval hints at purity, and is more likely to be considered distinctive than Irish Spring (86 percent versus 58 percent) but also triggers positive associations related to touch.

## *Getting Emotional*

If measuring the sensory response to brands was tricky, then measuring the emotional response is more so. Emotion is a construct with many flavors or tones, best described as follows:

"Emotion is a complex set of interactions among subjective and objective factors, mediated by neural/hormonal systems, which can (a) give rise to affective experiences such as feelings of arousal, pleasure/displeasure; (b) generate cognitive processes such as emotionally relevant perceptual effects, appraisals, labeling processes; (c) activate widespread physiological adjustments to the arousing conditions; and (d) lead to behavior that is often, but not always, expressive, goal-oriented, and adaptive."[4]

To put it more simply, emotions matter in marketing because they can help explain why people behave as they do and why they remain loyal to a brand. As Steve Heyer, former chief operating officer of Coca-Cola, stated of his flagship brand: "Coca-Cola is a feeling. Coca-Cola is refreshment and connection. Always has been . . . always will be."

It is beyond the scope of a simple survey to dig deep into the emotional response to the senses, but we can identify how the senses make people feel at a descriptive level. Associations with loved ones and childhood were common themes across many of the sensory associations people talked about in our research.

One respondent's reply to a question about taste illustrates the emotional component: "Sharing a Pepsi with my husband before we were married: two straws, one tall glass, and the tinkle of the ice as we drank. I fell in love with him that day."

In order to explore the emotional response evoked by the different senses, we start with the simplifying assumption that emotion is related to a good or bad stimulus response, in this case, the sensory impression people are describing. This reflects the basic human response to all stimuli, which is then interpreted by the individual into an emotional response based on context and mood.[5] We then classify the response using a series of word choices.

Many years ago, I took a personality test, the famous Myers-Briggs Type Indicator which is used extensively in business, education, and counseling for personal and professional growth. I remember being amazed at how well a series of simple word choices could define my personality (although the research conducted to ensure that word choices are discriminating and the analysis that makes sense of them are far from simple). This experience inspired my use of word choice questions for a number of different subjects, including measuring the degree of involvement for TV ads,[6] and now emotions. In the *BRAND sense* research we used three sets of six words each from which the person answering the survey had to select one word to describe how the sensation made them feel.

The word sets are based on an extensive literature review and subsequent experimentation and define six basic types of emotional response shown in Figure 6.6.

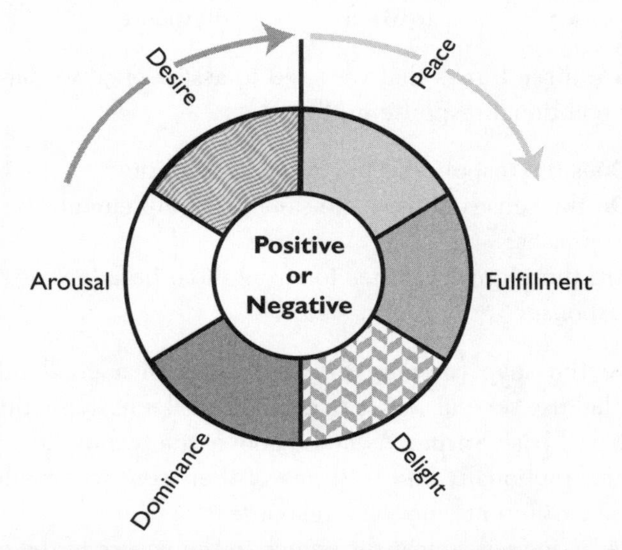


**FIGURE 6.6** The six basic emotional states are shown clockwise in order of activation from passive to active. Each state can be positive, attracting someone to the brand, or negative, repelling them.

Each response has a positive and a negative descriptor, defined by different words in the choice set:

Desire: the person feels attracted or repulsed.
Arousal: the person feels energized or irritated.
Dominance: the person feels empowered or inadequate.
Delight: the person feels happy or sad.
Fulfillment: the person feels contented or anxious.
Calm: the person feels peaceful or bored.

This is a very simple summary but it does capture most of the dimensions relevant to marketing.

## *Emotional Response*

When we come to measure emotional response to the brands included in our research, we need to take into account the saliency and nature of the basic response using the following formula:

$$\text{Emotional Response} = \text{Senses Brought to Mind} \times \text{Positive Response} \times \text{Emotion}$$

There are three things that we need to assess once we have identified the emotional response profile:

1. Does the response fit the brand positioning?
2. Do the senses deliver the same or complementary emotional responses?
3. Are there opportunities to strengthen, broaden, or refine the response?

Since the soap category provides us with a good example of brands that use several senses to good effect, let us apply this process to Dove and Irish Spring. As noted above, the two brands are different, both functionally and in terms of their sensory profile, but do they create different emotional responses?

There is indeed a clear difference in the nature of the emotional response. Dove users report that the smell of the brand leaves them feeling calm and satisfied. The other senses, particularly touch and sight, evoke similar responses, resulting in the overall profile shown in Figure 6.7.

Irish Spring is more likely to create a sense of stimulation and energy, consistent with the brand positioning. This is mostly driven by the brand's invigorating smell and is not as strongly supported by touch or sight. In this regard, I would argue that Dove delivers greater clarity of response than Irish Spring through complementary use of the senses. It would seem logical that if all the senses create the same positive emotional response, consistent with the brand positioning, the brand will be far stronger.

Dove users do claim to be more loyal to their brand and, as we see here, the sensory experience is tailored to create a greater depth of response. This is a sensible strategy for brands where the product and packaging are the main sensory stimuli. Retail brands, however, offer a far broader palette to work with, and some use the entire range of senses to create a highly desirable experience dependent on all five senses.

## The Starbucks Experience

The Starbucks coffee chain aims to engage the visitor with a multi-sensory orchestration to create an enjoyable and memorable experi-

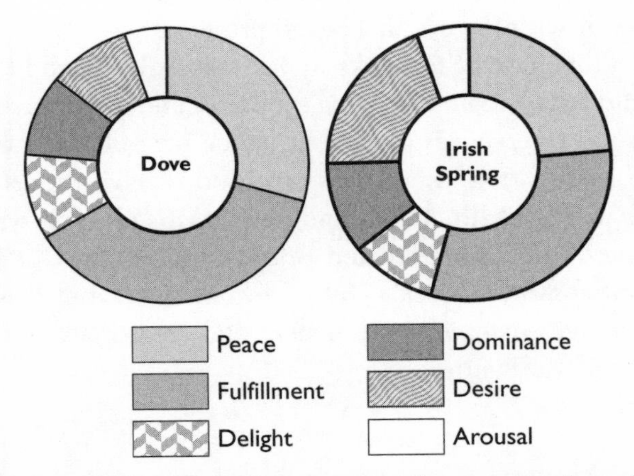


**FIGURE 6.7** Dove's emotional profile is more strongly concentrated on calm and fulfillment, whereas Irish Spring is more likely to create arousal.

ence that people want to repeat time after time. Howard Schultz's original vision of bringing the coffee bar culture of Milan back to Seattle has now morphed into its own culture, which can be found in over 7,000 stores around the world from Beijing to Santiago.

Starbucks' objective is to create an intimate and even romantic atmosphere where every element of the experience is tuned to distinguish it from other retail outlets. The lighting is subdued to keep it cozy. The chairs and floor may be a little scuffed, but that makes it feel comfy and homely, and less intimidating than lots of stainless steel. The action by the counter is clearly visible and provides visual entertainment. The music is soothing and sophisticated, merging into the background to create the right ambience.

Oh yes, and then there is the coffee. The enticing aroma permeates the entire location, creating anticipation of the rich roasted taste. Oh, and don't forget the Tazo tea, with unique flavors that revitalize the soul. Oh, and then we have Frappuccino and ice cream, to make us feel indulgent, whatever next? If Starbucks was just a cup of coffee rather than an experience, it would not be so easy to stretch the brand across geographies, locations, and products. I do wonder, however, if Starbucks has not gone a step too far with the newer airport locations. Do they fit the positioning well? Or do the bright lights, the noise of the announcements, the bustle of people running for planes combine to create a feeling of anxious restlessness completely at odds with the original brand promise?

If we had included Starbucks in our research I have little doubt that it would have created a strong, positive, and distinctive impression across all five senses, resulting in feelings of satisfaction and relaxation. In this case the senses are used to give the experience both depth and breadth. The brand's success is clear: it was recognized as one of the "Most Trusted Brands" by *Ad Week* in 2003 and ranked eighth on *Brandweek*'s "Super Brands List," but most important, the average customer visits a store eighteen times a month and spends over $60 a month.

## *Sensory Dilemma*

Appealing to the senses is by no means the only way to create a sense of leadership. Gillette has utilized product innovation to help it dominate the male shaving market for many years, culminating in the launch of the Mach 3 Turbo and the battery-powered M3 Turbo. The constant innovation and attendant publicity keep the brand front and center in people's minds. Over 50 percent of men are bonded to the Gillette brand in the United States and the UK. It is their first choice brand and they believe it beats the competition on all fronts.

Strong as it is, however, Gillette faces a basic problem, highlighted in Millward Brown's qualitative research. There is a fundamental dichotomy between the end benefit, a clean, smooth shave, and the delivery system, hard and, ouch! . . . sharp. Many brands in the shaving market, including Gillette, have used a common sensory cue in their TV advertising to try to highlight the efficacy of their product. We might call it "the touch test." Whether it is the shaver or someone else, running a hand over the recently shaved cheek or chin is designed to signal smoothness. But then we see the three blades being bonded into the razor. It looks hard, shiny, and definitely sharp. And which one do you think evokes the most memories? Nobody ever stroked my cheek and told me how smooth it was, but I have certainly nicked myself shaving enough times! Does the advertising achieve clarity, or is the response somewhat diluted by the implied sensory conflict?

If the Gillette advertising might send people mixed messages, however, what of the advertising for Remington in the run up to the 2003 holiday season in the United States? Here we see an exotic sports car being sliced in half by a giant band saw in order to demonstrate the sharpness of titanium blades. The saw rips through the car, narrowly missing the driver. Yes, it conveys a feeling of power and efficacy, but what other sensory impressions and emotions might it evoke?

## *Highlighting the Sensory Experience*

Dove soap and Starbucks provide great examples using the senses to complement each other and maximize the emotional differentiation of a brand. Together the senses combine to create a compelling experience that supports the brand's unique positioning. Clearly this did not happen by accident. Considerable time and effort will have been devoted to creating these experiences, but is that enough to guarantee success?

As we saw from the home electronics example, consumers can remain stubbornly unaware of some aspects of the brand experience. Years of home electronics marketing have focused on the primary benefits of picture and sound quality, supported by long lists of technical specifications that few people really understand. Only a few premium brands, like Bang & Olufsen, choose to highlight design and feel as well as the primary senses of sight and sound. Consumers are well versed in the language of what makes a good picture (clear, bright, and with good definition), and what makes a great sound (clear, rich, and with good definition), but there are no cues that encourage them to go beyond these two benefits when considering the merits of different brands. They will ultimately do so in the store, but by then it is purely down to the appeal of the design of the day; no sensory cues have been established ahead of time to sway the decision one way of the other.

If there is no readily identifiable criterion for making a choice between brands, then consumers are unlikely to find one on their own. So what the smart marketer needs to do is highlight areas of competitive advantage and make sure that the idea is firmly fixed in the minds of consumers. This is what my brand stands for, this is why you will enjoy it. The positioning of Dove and Irish Spring does not leave expectations or experience to chance. Both brands invest heavily in above-the-line advertising in order to establish the brand promise and to frame the usage experience to ensure a positive response from consumers. This point was driven home to me when I checked out the Irish Spring website. Although the look of the site was not particularly stimulating, featuring pastoral scenes, the lively

Irish jig that played on the home page was entirely in line with the sense of invigoration the brand creates for its users. It is a great example of how a sense not normally salient for the brand can still be used to evoke an emotional response entirely consistent with the brand positioning and experience.

## *The Metaphorical Sense*

Across all of the brands we surveyed, sight tended to play a supporting role to the other senses. The research demonstrates that taste, touch, and smell are intimately involved in creating a great brand experience and continued loyalty. The role of sound to create an emotional response is well documented.[7] Sight may convey information well, but even at best it creates a less deeply felt emotional response. This presents marketers with a challenge: how do you communicate the brand experience if you cannot use sight?

Well, if you cannot lure your consumers to the brand, you can take the brand to the consumers. In the UK Audi used an insert in *What Car?* magazine to focus attention on the new-car smell for the A6. The insert was a piece of interior leather from the A6, shaped like the car, and inscribed with "Experience the smell of the Audi A6, now with interior leather as standard." Comfort Fabric Conditioner adopted a similar strategy using a direct mail campaign of sending out swatches of cloth impregnated with their new fragrance.

Tactics like these, however, are expensive and ignore an important loyalty-building device. By taking the brand to the consumer you miss out on the ability to create expectations that consumers can then confirm for themselves. Although sight might be the weaker sense in terms of creating a strong emotional response, it can frame people's expectations of the usage experience. If you visit the Pepsi website you will find a large, but empty, glass. As the page loads, ice cubes rattle into the glass, Pepsi pours over them accompanied by the familiar fizzing sound, and dew forms on the outside of the glass. Refreshing drink anyone? This enhancement of the usage experience is an important means by which marketing communications can create brand loyalty.

A classic example of creating expectations, largely through sight but augmented by sound, was the relaunch advertising for Clairol Herbal Essences. The shampoo category is characterized by constant innovation. New brands appear on a regular basis, and in order to stay the course, established brands must find new ways to reinvent the brand in the mind of the consumer while staying loyal to what they stand for. Clairol had been very successful in the 1970s, but by the early '90s that success was wearing thin. Clairol relaunched the brand, holding true to the organic and natural positioning, but using a unique advertising approach to highlight the new herbal ingredients. Unique because it abandoned the traditional strategy of advertising the end benefit—clean, glossy hair—in favor of playing up the sensory experience of using the shampoo. The TV advertising took this to the extreme by playing on the fake orgasm scene in the movie *When Harry Met Sally* . . . . The experience of showering with Clairol Herbal Essences is vividly portrayed on screen; there is no doubt that the woman in the shower is having a very good time. And another woman in the ad, watching this performance on her TV, is in no doubt about it. She turns to her partner and says, "I want the shampoo she's using."

The advertising was very successful and helped revitalize the brand as well as its consumers. This success, however, was not just due to the notoriety created by the commercial. Any person who has used the brand will know that each of the varieties has a strong, pleasant smell. Released by the hot water, this permeates the air, working in synergy with the feeling of the shower itself to make the user feel awake and invigorated. The user may not feel the way the woman in the commercial is purported to feel, but she certainly does notice the fragrance and respond positively to the experience. Even in the highly competitive shampoo market, Clairol Herbal Essences remains a strong brand today.

## Working with the Senses

All of the brands we researched for *BRAND sense* had at least one distinctive and positive sensory attribute, and some had several. A few of the brands, however, failed to leverage these properties to their full effect, a situation likely to be true of many brands. Some brands were even associated with sensory impressions that stimulated a negative response, clearly weakening the overall proposition.

Those of us who work in the world of marketing spend a lot of time trying to find new ways to differentiate brands. Like the people who buy those brands, we sometimes lose sight of what is under our noses, at our fingertips, or even in plain sight. One of the most compelling ways of ensuring competitive advantage is through the senses. They offer more than a means to satisfy basic needs. They offer an opportunity to confirm and enhance the brand promise and create a strong emotional bond with consumers.

Over thousands of years religion has appealed to our five senses in fascinating ways. Controversial as it may seem, in chapter 7 we see how it may very well provide the inspiration for future sensory brands.

---

### Highlights

The *BRAND sense* study is designed in such a way that it makes respondents think about their own senses and which ones they are most sensitive to. It covers the differences between brands based on the senses before finally getting to the core of the questionnaire focusing on the senses and specific brands.

The study revealed that all the brands researched for *BRAND sense* had at least one distinctive and positive sensory attribute, some had several. Some of the brands in the study failed to leverage these properties to their full effect, a situation that's true for many brands. Some brands were associated with sensory impressions that stimulated a negative response, clearly weakening the overall proposition.

The study reveals that sight does not have a significant relationship with experience across the brands, but wields influence through creating perceptions of leadership and clarity. Across all of the brands surveyed, sight tended to play a supporting role to the other senses. The research demonstrates that taste, touch and smell are intimately involved in creating a great brand experience and continued loyalty.

Even though there was relatively little difference in consumer loyalty between brands like Coke and Pepsi, the Millward Brown research suggests that the senses do have a role in creating competitive advantage in the cola category. The same is the case for almost any other category.

---

## Action Points

1. Think about your brand and make a list of the sensory impressions it brings to mind. Then identify whether each impression is positive or negative.
2. For each sense on the list, identify the primary emotion that it might evoke.
3. If there are some significant negatives, how are you going to eliminate them? If you cannot remove them, how can you tone them down or divert attention from them?
4. Do the positive emotions work in harmony to support the brand positioning or are they discordant? How could you bring them more into line? What can you do to strengthen the emotional response using synergy between the senses?
5. The emotional landscape changes over time. The 1990s were the "me" years, when dominance and arousal were the primary emotions. Brands that implied status and indulgence prospered. The events of 9/11 brought on a period when fear was a dominant emotion. Brands that offered reassurance and calm prospered for a while. Think about the likely trends of the future. What might your brand be able to offer that strikes a chord with those trends?
6. If you think you have identified problems or opportunities, check them out by conducting research with your brand's consumers to confirm your hypotheses. Check out your competition at the same time. There may well be issues that you can exploit or advantages that you can undermine by adapting your brand's sensory profile.
7. You are not there yet. Religions have leveraged the five senses for thousands of years. Your brand can most likely be inspired by this. Chapter 7 reveals how.

CHAPTER 7

# Brand Religion: Lessons Learned

**DAVID LEVINE, CHRISTOF KOCH, AND** Mark Tappert are three men who live thousands of miles apart. There's a fifteen-year spread in their ages and they have totally different careers. David is a psychologist, Christof's a professor of computation and neural systems, and Mark works as a graphic designer. Yet despite these disparate demographics they have one important thing in common. Tattooed on their right arms is an apple. Not just any old apple, but a very specific one. The one with a clean bite and a short stalk. The globally recognized symbol of Apple computers.

To permanently etch the apple logo into their skin is a sign of their unwavering faith in the Apple brand. For these three men, Apple has become an addiction like the faith and loyalty someone feels to one's favorite sports team or music group. It could even be likened to religious fervor.

Religious fervor is primarily based on faith and belief. At the risk of sounding crass, these days it is sometimes hard to divorce faith and belief from big business. According to *Wired* (December 2003), religion racked up $3.6 billion spread across various media, from bibles to incense, candles to psalm books, in 2003 alone. By contrast, video games accounted for another $200 million and a further $2.5 billion was generated from nonreligious book sales. There's no doubt that this is a reflection of the somewhat uncertain times. Wars, financial challenges, a changing labor market, more people, fewer jobs, escalating crime, and an increasing divorce rate—as these and other uncertainties fill our days, so there's an increasing need for stability. Only the unwise consumer would invest time or money in anything that he suspected would not survive. Instead, there's a conscious seeking for permanent foundations that offer solid promises. For a large percentage of people, religion fits the bill in a world that's changing at an incomprehensible pace. It offers lifelong guidance on how to live, and can provide a road map that extends way into the future, even assuring some of security beyond death.

Online games take people to other worlds where they meet fellow travelers, play together, test their skills against one another. They offer the player total control. There need be no arguments or unhappy situations because retreat from anything unpleasant is a mere click of the mouse away.

At first glance, branding and religion are an odd, incongruous combination. But on further examination the relationship is closer than we dare imagine.

Branding continuously strives to achieve authenticity and build a relationship with consumers that will extend from cradle to grave. By its very longevity, religion automatically assumes an authentic, loyal, lifelong relationship with its adherents. Brands attach labels to physical products or services, whereas religion represents intangibles—phenomena difficult to describe and impossible to show. Brands are constantly locked in value-based discussion. Religion stays in tune with the ephemeral, a place that marketers can only dream of.

David's, Christof's, and Mark's devotion to Apple is somewhat unique, although it does beg the question as to why so few brands have managed to create such a devotional fervor. Is it reasonable to think that marketers can learn much from religion when they are thinking in branding terms?

## *Brand Loyalty*

The more loyalty a brand inspires, the greater the potential it has for long-term success. In fact, loyalty is a primary factor in creating this type of success. As in all matters to do with belief, you can't weigh, predict, or purchase loyalty. It is a result of a range of trust-generating factors, which over time builds the kind of allegiance every marketer aims for—the loyal customer.

Loyalty is strong, but traditions are even stronger. Tradition is partly formed by long-term loyalties that have become so ingrained within a culture that rational behavior has long since given way to emotional affiliations. Even if some brands claim a strong loyal following, none can claim to be part of a tradition.

However, we're surrounded by traditions of both the secular and sacred variety. We pop champagne corks and watch fireworks on New Year's Eve. Hindu homes are lit up for two weeks each year to ward off dark spirits during Diwali. Florists more often than not run out of red roses on Valentine's Day, and once a year on the Day of Atonement Jews fast to cleanse themselves of sin. Many of these traditions are accompanied by gift giving and special foods.

We pay the hiked-up prices for flowers on Valentine's Day and find time in busy schedules to send Christmas cards because these rituals are woven into the fabric of our lives. We do all this despite the fact that so much of it does not stand up to rational scrutiny in the twenty-first century. We barely question it, we just do it. We do it because traditions and ritual offer a sense of predictability. The actions bind us to our community and generally make us feel secure and comfortable in our lives.

Traditional celebrations—whether thought up by clever marketers or evolved over centuries—are centered around rituals. In addition, families have a tendency to create their own private rituals. Whether we're served green cake on our birthdays or get married in Grandma's wedding dress, we participate in these irrational customs because they give us a sense of predictability and belonging in a world where everything is changing.

As I've briefly discussed in chapter 5, we each develop our set of daily rituals too. Yours may be to grab your double espresso from Starbucks every morning on your way to work, and end your day with a Bud. For many, movies just aren't movies without a Coke and popcorn to accompany them. Notice how these brands have worked their way into the lexicon. They have in fact managed to take their followers up a rung on the loyalty ladder to become part of tradition.

As in religious traditions, branded tradition can be passed from generation to generation. Some families make annual visits to Disneyland. During the long month of Ramadan, when Saudi Arabians fast all day, as many as 12 percent of the population wash down their evening meal with Sunkist, a concentrated juice product. Sunkist has become part of their ceremonial rituals.

## *Superstitious Bonding*

As we continue up the loyalty ladder toward brand heaven there are very few brands that have achieved superstitious bonding—they've become a way of life, almost not seen as commercial products any longer.

Jack Nicholson won an Oscar for his role of Melvin Udall in the movie *As Good as It Gets*. He played a curmudgeonly man with an obsessive-compulsive disorder. Walking across checkerboard floor tiles demanded painstaking concentration to ensure that he never stepped on the black squares, only the white. Audiences the world over chuckled at this idiosyncrasy. We laughed because some part of us recognized and identified with it. We "touch wood" to prevent a horror we've spoken of becoming reality, as if it will change the course of destiny. We don't walk under ladders, and on Friday the thirteenth we remain slightly more vigilant. We all engage in some form of these little practices to make our world safer. Is it rational? Hardly. However, we fear consequences if we don't. Superstition becomes tradition.

The Beatles' "Rocky Raccoon" checked into his room where all he found was a Gideon Bible. Rocky is not alone. In hotel rooms from Liverpool to Lahore, tucked away in the bedside table drawer is a

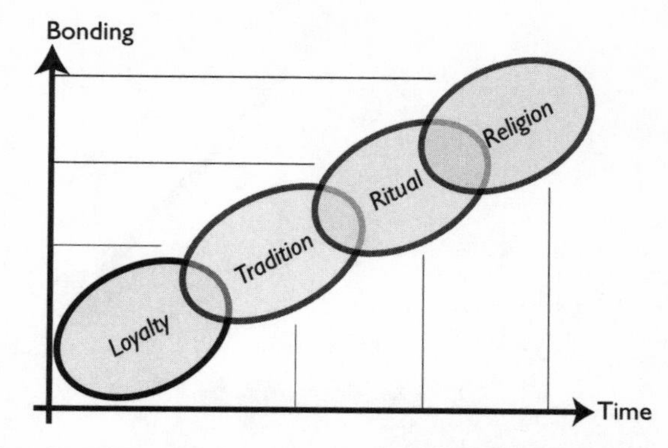


**FIGURE 7.1** Loyalty is the starting point for branding. Over the next decade brands will explore ways to substantially improve the strength of their bonding with customers.

religious book—a Bible or a Koran, perhaps. There's no law anywhere that prescribes this, yet for over one hundred years, travelers in hotel rooms from Sacramento to Sydney can rest easier knowing spiritual sustenance is but a drawer away.

No brand has yet succeeded in achieving such a level of dependence or trust, nor should we expect it to, but religion does provide a role model in terms of offering wisdom and depth of meaning. Brand builders can learn from the way religion has communicated its message through myth, symbol, and metaphor over the millennia. We're absorbed by the stories and captivated by the history, symbols, and historical messages. They touch us at a fundamental emotional level, which precludes any rational discussion.

In contrast, branding has become an ever more rational science. We focus almost exclusively on short-term financial outcomes. As a result, brands have rejected the emotional propositions put forth in the 1950s and 1960s in their price-driven quest to create value-added ties to the consumer. We need to reframe our thinking. Perhaps it's time to swing back?

There's no doubt that people are searching for emotional fulfillment. So much focus is now on rational argument and measurable outcomes that there's a growing need for emotional connection. The steady attraction to alternative religions has become a fact of life. Research undertaken with the tween generation (eight-to-fourteen-

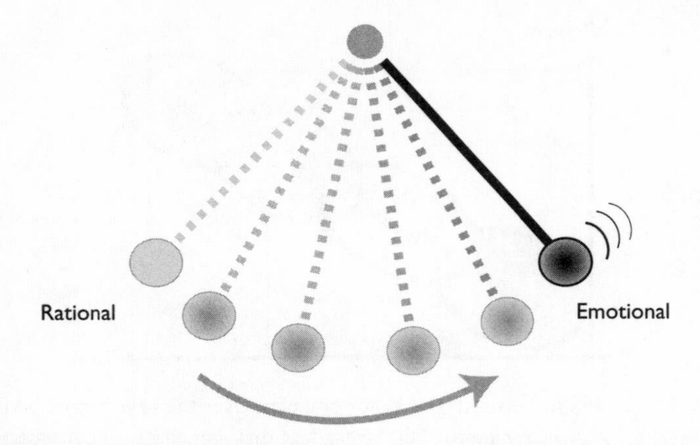


**FIGURE 7.2** Over the past 100 years, a constant shift has taken place between emotional-focused branding and rational-focused branding.

year-olds) for Project BRANDchild demonstrated a clear preference for emotions to be included in their entertainment, their commercials, and their brands. In fact, 76.6 percent of the tweens interviewed in the United States said they wanted something to believe in. A further 83.3 percent of all urban tweens regarded "obeying rules" as one of the most important phrases in their life. In fact, this figure was constant across the board, regardless of what country they were born and raised in. Given this intense need for emotional connection, it's not surprising that between 1991 and 1997 sales of mainstream religious and spiritual books to the general population increased by 150 percent, compared to a 35 percent increase in sales for secular books. [1]

What becomes apparent is that the gap between the rational proposition of today's brands and the need for emotionally satisfying products, services, and beliefs is growing. This is a need that's properly being met by religion. If for no other reason, and without equating the two or seeming overly exploitive, branding should aspire to build ever-stronger emotional ties with its customers.

## *Religion as Brand*

### Icons

The best-known icons in the world are religious: the cross, the crescent, the meditative Buddha, the om, the ankh, and the Star of David. Each carries an enormous symbolic weight. Each represents a way of life, a belief system, and a community for millions, if not billions, of people. Despite myriads of graphic representations of these icons that proliferate throughout the world and continue to be created, these icons are easily recognized. Apart from their straightforward message, we've seen them morph into all kinds of permutations. They're set in pendants, on flags, in art, atop buildings, on T-shirts, and in print. They come in every material substance you can think of.

Royal families the world over are represented by symbols or logos. Each member of the royal family in Denmark has his or her own monogram. A new monogram is created when a new member enters the family and is designed to reflect the member's personality.

Iconic messages can be multilayered and ambiguous. Buddhas have been adopted by a multitude of non-Buddhists of other faiths who are seeking enlightenment in one form or another. Decorative crosses have become fashion statements for styles as diverse as the

**FIGURE 7.3** In the finest fairy-tale fashion, the Prince of Denmark married Mary Donaldson, an Australian commoner, on May 14, 2004. The monogram that she shared with her husband, HRH, The Crown Prince, was designed by his mother, HM Queen Margrethe II.

wayward gypsy style and the hard-core studded-leather punk look. Religious icons are the equivalent of the marketer's ultimate branding logo, except that marketers are looking for longstanding tradition as well as for instant and widespread recognition. They don't want to have to wait a couple of thousand years to get there.

## Spreading the Word

There was a time, not that long ago, when nautical maps of Europe had legends that included churches on the land. You often could find your way by seeing those spires before you happened upon a lighthouse, and so the churches became an important navigation tool.

Churches once held the monopoly on the choice real estate of the town, and they tended to pick the highest ground, so their grand steeples and graceful spires topped with a cross dominated the horizon. No buildings were allowed to be built higher, ensuring that the church would always occupy the place closest to heaven. Despite modernization, the city of Rome still abides by a law that states that no building can be higher than the dome of St. Peter's.

Christianity is not alone when it comes to seeking maximum visibility for its places of worship. Mosques and minarets dominate the skyline in many Muslim cities where crescents shine from the top of the towering minarets, and amid the ever-changing high-rise horizon of Bangkok in Thailand, the tall curved roofs of shining gold Buddhist temples remain easy to spot.

So all over the world religious buildings are not only visible, they're accessible from every direction. The whole town knows where they are. They're central to the formation of community, providing a sense of belonging and a sharing of core values.

## Follow the Leaders

If you visit Bangkok's Pariwas temple in Thailand you will find more than the usual Buddha. A fan of soccer star David Beckham has taken celebrity worship to a new level. A one-foot-high shimmering gold-leaf statue of Beckham takes its place at the feet of Buddha along with other minor deities. Chan Theerapunyo, the temple's senior monk,

**FIGURE 7.4** A sharp eye would notice that one of the 50 Buddhas at Pariwas Temple in Bangkok is slightly different. It's a Beckham Buddha—in gold.

has defended this, saying, "Football has become a religion and has millions of followers. So to be up-to-date, we have to open our minds and share the feelings of millions of people who admire Beckham."

The adoration of David Beckham pales in comparison to a Japanese cartoon character called Hello Kitty. Over twenty-five years this bulbous cat with no mouth has earned the Sanrio corporation literally billions. A website called Praying for Hello Kitty gives an indication of the power of the brand. This small pale character known as Hello Kitty has become an almost religious brand in Japan. It's hard to dispute the underlying religiosity of a message posted on the Praying for Hello Kitty website that includes the lines "Hello Kitty is a white angel who does not know any dirty. Hello Kitty is the saint Maria . . . Hello Kitty is the creature which the God made first . . . Hello Kittyist world will prosper more and more. Jesus Hello Kitty. Our Hello Kitty . . ."[2] This is but one example that indicates the power of the brand.

Just as each religion is built around strong charismatic leadership, so these qualities are reflected in our most successful personality brands. The David Beckhams, Bonos, and Madonnas of the world demonstrate a similar power with armies of devoted followers. More traditional brands follow exactly the same trend because of their strong charismatic leaders. Think Richard Branson, Walt Disney, Steve Jobs. Each name has become synonymous with the brand he has built. The founder is the brand, and the brand—be it Virgin, Disney, or Apple—is the definitive light in their lives. It's a light that can sometimes dim. Steve Jobs experienced his moments in the wilderness when he was ousted from Apple for a short while. Martha Stewart, the doyenne of household brands, has experienced her own fall from grace.

At a Macromedia conference in San Francisco some years ago I sat beside an Apple devotee who held a Newton. This was a PDA introduced by Apple in the mid-1990s that despite its conceptual genius never managed a successful breakthrough. Steve Jobs was a surprise speaker. Dressed in his characteristically casual attire, he began his speech by proclaiming that the Newton was dead and dramatically throwing one of the devices into an Apple bin on the stage.

Some people intuitively applauded, some were screaming, and others were crying. After the person next to me spent some time furiously taking notes, he joined the fracas, threw his Newton on the floor, and started to jump on it. For him Apple was more than just an electronics manufacturer—it was closer to a religion.

In another famous corporate marketing failure, the formula of Coca-Cola was changed. In 1985 Coke decided to follow the results of a taste test after a survey of nearly 200,000 consumers. The New Coke marked the first change to the secret recipe in ninety-nine years. When the taste change was announced in 1984, some consumers panicked and filled their basements with cases of the original Coke.

The results were catastrophic. Consumers were outraged. They felt that a small piece of Americana was being discarded. Calls flooded in on the 800-GET-COKE phone line, as well as to Coca-Cola offices across the United States. By June 1985, the Coca-Cola Company was receiving 1,500 calls a day on its consumer hotline, compared with the average 400 before the taste change.

What the survey's taste tests hadn't shown, of course, was the bond consumers felt with the Coca-Cola brand. In this quasi-religious relationship, consumers didn't want anyone, including the Coca-Cola Company, tampering with the brand. Protest groups like the Society for the Preservation of the Real Thing and Old Cola Drinkers of America (which claimed to have recruited 100,000 members in a drive to bring back "old" Coke) popped up around the country. Songs were written to honor the old taste. Protesters at a Coca-Cola event in downtown Atlanta in May 1985 carried signs proclaiming "We want the real thing" and "Our children will never know refreshment."

The return of original formula Coca-Cola on July 11, 1985 put the cap on seventy-nine days that revolutionized the soft-drink industry. The transformed Coca-Cola Company stands today as testimony to the power of brands that are more than just brands. Consumers applauded the decision. Unbeknownst to the Coca-Cola Company, Coke had taken on the trappings of a mini religion.

## *Inspired by Religion*

Harley-Davidson, Apple, and Coca-Cola have all provoked reactions and actions from their customers rarely experienced by other brands. These customers are more than just customers, they've been converted into full-fledged believers. You could argue that in their devotion, any aspect of rational thinking is absent.

So, what does it require to take a brand beyond its traditional loyal base of consumers toward a bonding that resembles a religious relationship? The first step requires paying close attention to the Ten Rules of sensory branding. These are the fundamental components that underpin religion and can serve as the ultimate role model for branding.

1. A unique sense of belonging
2. A clear vision with a sense of purpose
3. Take power from your enemies
4. Authenticity
5. Consistency

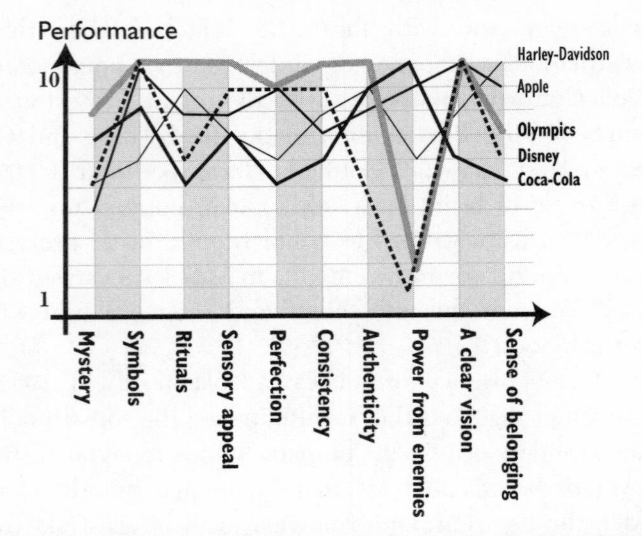


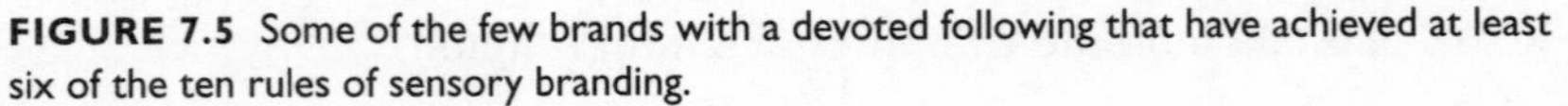

**FIGURE 7.5** Some of the few brands with a devoted following that have achieved at least six of the ten rules of sensory branding.

6. Perfection
7. Sensory appeal
8. Rituals
9. Symbols
10. Mystery

## 1. A Unique Sense of Belonging

Every religion fosters a binding sense of community. Within the bosom of this community belief can grow, be stimulated, and cement relationships among its members, creating powerful feelings of belonging. Living in the same neighborhood or sharing a common culture does not necessarily create this sense. There has to be a social glue that binds common goals and values, and celebrates and mourns the same events. As the congregation's members invest their time and resources in the community, they create social capital that further enhances the sense of belonging.

Do customers using your brand feel that same sense of belonging? Say, for example, your brand encountered a crisis. Would your consumers establish networks to support the brand? Not possible? Well, this is exactly what happened when Michael Jackson was again charged with child molestation in November 2003. The Michael Jackson Fan Club established Fan Watch with the clear objective of raising money to ensure Jackson receives fair treatment in both the media and the judicial system. It is rumored that over $2 million was collected in just two months.

Music communities have existed for as long as there's been music. But since the advent of the Internet, they've become well-oiled information disseminators and are now emulated by companies that seek to get their particular artist across. It works for music sales, so it follows that brand clubs might perform in the same way.

Let's revisit David Levine, the self-professed "Mac nut" and lecturer in psychology at the University of Illinois. Levine is the proud owner of a suede Apple jacket adorned with some of the original Macintosh icons. He paid $400 for it, but has yet to wear it outside of home. In fact, his household is a mini Macworld. He owns ten of them. He recently spent $4,000 on a dual-processor G4, and another $2,000 on the large flat-panel cinema display. He says, "I don't need it. I did it to support the Mac."

Like all Apple collectors, Levine identifies strongly with Mac culture. He also owns Apple-logo T-shirts and even Apple-brands the tags on his luggage. He acknowledges belonging to a Mac community and acknowledges the religious connotations of his Apple affiliation. He believes that Mac users have a common way of thinking and of doing things. They share a certain mind-set. "People say they are a Buddhist or a Catholic," he added. "We say we're Mac users, and that means we have similar values."

Establishing a sense of belonging is fundamental to growing a community. There are approximately 2 million online communities on the Internet, although curiously less than 0.1 percent of them are specifically related to brands. Perhaps this reflects the reality that almost all brands fail to place the user at the center of communication. Consumers get tired of talking about a product day in and day out, but they will latch onto individual aspects that unite them with other users forming a common bond. The role of the brand therefore is simply to create the social glue to connect people, finding their common ground and thus, most importantly, generating a strong sense of belonging.

One of the largest branded communities in the world is Weight Watchers. More than 2 million members conduct 30,000 meetings a year across twenty-nine countries. Weight Watchers has worked on bonding people with common weight challenges for over forty years. Their meetings offer encouragement and advice to help their members achieve their weight-loss goals. It is one of the few brands based on the concept of generating a sense of acceptance and belonging. Then they take their community one step further and provide a workable strategy to slim down.

Another brand that fosters a sense of belonging is the construction toy LEGO. There are 4,000 LEGO communities made up of every age group around the world. So if you're a grandfather of seventy-five who loves building with LEGO blocks, no one in this community would look at you askance. They share the passion, and thereby legitimize it. A quick exploration of the LEGO community reveals a diverse array of members, from mathematics professors to the unemployed.

Then there are the devoted Manchester United aficionados who make up some 10,000 different communities spanning from individual football fans to those who play on the team and have since become heroes in Japan.

Everyone, everywhere feels the need to belong.

## 2. A Sense of Purpose

Christof Koch is one of the world's leading neuroscientists, teaming up with Nobel Prize winner Francis Crick to search for the neurological seat of consciousness. Koch had his Apple tattoo done while he was on an archaelogical dig in Israel.

What motivates supporters to etch the Apple logo onto their skin, or Michael Jackson fans to pay hard-earned pocket money for his legal defense? Well, these brands reflect a strong sense of purpose. Their followers are more than devoted fans, they're positively evangelical in support of their chosen brand. Some brands are represented by visible, daring, or determined leaders. Richard Branson of Virgin has made sev-

| TATOO BRANDS | percent |
|---|---|
| Harley-Davidson | 18.9 |
| Disney | 14.8 |
| Coca-Cola | 7.7 |
| Google | 6.6 |
| Pepsi | 6.1 |
| Rolex | 5.6 |
| Nike | 4.6 |
| Adidas | 3.1 |
| Absolut Vodka | 2.6 |
| Nintendo | 1.5 |

**FIGURE 7.6** How do you measure the ultimate brand loyalty? Perhaps by asking consumers what brand if any they would be prepared to tattoo onto their arms. Interpret the somewhat surprising results by yourself.

eral well-publicized attempts to pilot the first solo flight around the world in an air balloon. Steve Jobs returned to an ailing Apple and in less than a year managed to turn around the whole company's fortunes for the grand salary of $1. This is the stuff that legends are made of. Daring stunts and headline-grabbing leaders will not succeed unless the vision of the brand has a strong consumer-focused foundation. Koch wholeheartedly identifies with Apple's strong vision. Apple exhorts its customers to "Think different" and aligns itself with such luminaries as Albert Einstein, John Lennon, and Mahatma Gandhi. Their message points out that at the core of the Apple brand is a philosophy that guides its users which is far deeper than its stunningly stylish technology.

A brand needs to set challenges, question them, then conquer the challenges by turning itself into a hero—in just the same way that musicians, sport stars, and movie celebrities do. And at the root of every challenge there must be a clear sense of purpose that will help consumers identify exactly who they are in relation to the brand.

## 3. Take Power from Your Competitors

When recalling the cola battles in the 1970s, a senior executive at the Coca-Cola Company said, "Going to work was like going to war."

The challenge between Coke and Pepsi took on a global dimension that has slight echoes of centuries-old religious conflicts. The Bible or the Koran? Protestants or Catholics?

And war—in all its dimensions—is what brand warfare is all about. It unites citizens of large diverse nations with a single purpose. There's camaraderie of team or nation during the U.S. Super Bowl or the World Cup in soccer. And whose side are you on in the Microsoft-Apple debate? Avis swooped on a market by declaring itself Number Two, but at the same time saying, "We try harder." They've used the same slogan for forty years, constantly assessing their status as Number Two—an interesting proposition considering the general desire to be Number One.

Pierre Bourdieu, a French psychologist, once said, "A choice of brand is a clear statement of who you are not!" A visible enemy gives people an opportunity to show their colors and align themselves with the team or player that they most strongly identify with—and that may be the underdog. In 1991, Linus Torvalds, a twenty-one-year-old student in Finland, developed a new computer-operating system called Linux. Linux has since become hugely popular. Research indicates that about a third of all Web servers in the world are powered by Linux. That makes it the second most used system after the all-powerful Microsoft.[3] Linux has the distinction of being Microsoft's only competitor that has, to date, managed to compete successfully with the software giant. Linux followers' passion for the product is legendary. They have a revolutionary zeal to build a reliable product for the people. Many a Linux fan will sport a Linux tattoo, but you would be hard-pressed to find a Microsoft logo etched into anyone's flesh.

To become strong, a brand almost needs to be positioned in relation to another. There must be contrast and conflict in order to build up an "us" and "them" situation. In the 1980s, the ice-cream business in the United States was controlled by three multinational corporations. So when Pillsbury, the owner of Häagen-Dazs, wanted to limit the distribution of a small Vermont ice-cream company, Ben & Jerry's, the young ice-cream makers responded with its now famous "What's the Doughboy Afraid Of?" campaign. This campaign proved so successful that their sales in one year increased by 120 percent.

The David versus Goliath scenario gained consumers' empathy and support and helped place Ben & Jerry's product firmly on the supermarket shelf. Over the next decade Ben & Jerry's grew to become one of the biggest players in the U.S. ice-cream market. This ultimately culminated in Ben & Jerry's takeover on April 12, 2000 by Unilever, an Anglo-Dutch corporation, ending a twenty-two-year battle for the consumers' hearts and minds. David had become Goliath. The fight was over.

Ironically, very few companies leverage this technique. Most opt to pretend their competition doesn't exist. But as in any top game, movie, sports event, or political campaign, it's the tension that the competition presents that generates the excitement and involvement. It creates fans and enemies alike. There's passion and energy, opinion and argument. Everything that one can hope for when building a brand.

## 4. Authenticity

Authenticity is an essential component of any religion's history, anecdote, and mythology, and religious fervor rarely exists without a large dose of it. The world's major religions have established their credentials over the long course of time—often thousands of years. Antique stores the world over carry artifacts that often have religious connotations reflecting worship, heritage, and authenticity combined in one object.

But what is authenticity? The Merriam-Webster dictionary defines it as "worthy of acceptance or belief as conforming to or based on fact." It also means "conforming to an original so as to reproduce essential features," as well as "true to one's own personality, spirit, or character."

In a world of pirated copies of clothing brands, music CDs, and general merchandise, it is sometimes difficult to tell the authentic from the fake. Sydney took matters into their own hands at the 2000 Olympic Games. They introduced Brand DNA. Authentic DNA was extracted from the blood of selected athletes, replicated, and included as an ingredient in the ink used on all official merchandise carrying the Olympic logo. Armed with specialized equipment, officials were able instantly to determine the real stuff. This resulted in the destruction of truckloads of merchandise that didn't pass the Brand DNA test.

Unfortunately not every brand comes equipped with an authentication meter. Judgment is simply left to the public to determine. A brand's history goes a long way to help generate perceptions of authenticity. Photographers are generally aware that it was a Hasselblad camera that took the first pictures on the moon. By then the Swedish company had been manufacturing their precision equipment for over one hundred years. When Arvid Hasselblad established a photographic division in the company his father founded, he is reported to have said, "I certainly don't think that we will earn much money on this, but at least it will allow us to take pictures for free!"

Hasselblad cameras remain a brand of choice for both those in the profession and keen amateurs. Not only do they pass the most stringent quality tests, but they're well ahead on the authenticity scale. Hasselblad cameras have what the Japanese refer to as *miryokuteki hinshitsu*—that is, above and beyond the expected quality, they have a quality that fascinates. It's this elusive quality that fascinates that today's consumers demand. They feel cheated if they're merely presented with *aterimae hinshitsu,* which roughly means "you get what you pay for."

Manchester United is a soccer club that has *miryokuteki hinshitsu* in spades. Almost any one of its 53 million supporters can relate its story. After the Second World War the club was bankrupt. Matt Busby was brought in as manager, and in less than a decade he miraculously turned their fortunes around. By 1956 Machester United qualified to enter European competition. They reached the semifinals of the European Cup. A year later they took home the league title. But in the winter of 1958 they went to Belgrade to play against Red Star. On the journey home their plane stopped to refuel in Munich. It was snowing heavily and the runway was icy. There were two aborted takeoffs, and on the third attempt the plane overshot the runway and its wings clipped a house. The crash was responsible for the deaths of twenty-three people. Eight of them were from Manchester United's young team, known as "Busby's Babes."

The club's ability to overcome the tragedy is the stuff legends are made of. Their resilience and perseverance have gone a long way to creating the strength of the Manchester United brand, which earned $233 million in 2002. The irony of the situation is that even though they haven't been the world's best-performing team, their supporters have remained steadfast and their legion of fans continues to grow.

History adds the necessary credibility to brands. It supports notions of authenticity, which is one of the reasons why a brand's background and the stories surrounding its image become increasingly important. The survival of the Manchester United team or the Hasselblad camera generates trust and adds authenticity to the brand. The question then becomes, how do you succeed in creating authenticity if your brand has not stood the test of time?

Tim Hortons, the Canadian chain with the good cutlery, found a way around the authenticity problem for their new coffee outlets. They drew on the support of their customers, and based their entire advertising campaign on their testimonials proclaiming it a "meeting place—a home away from home." They emphasized the distinct Canadian-ness of the fresh homemade baked goods. They also created myth. One of the brand's key symbols, the coffee mug, became the focus of the story. As the tale goes, a Canadian who had traveled halfway around the world was approached by two Canadians who recognized him as a fellow countryman by his Tim Hortons mug. This formed the basis for a lifelong friendship.

## 5. Consistency

In addition to authenticity, Tim Hortons capitalizes on its stability. In an increasingly unstable world, this goes a long way. When you visit a Tim Hortons outlet, you know what to expect. You can count on things being "Always Fresh." It's entirely predictable in much the same way as a visit to your temple, church, or mosque is. It's always there when you need it. The more it stays the same, the more your belief is reinforced. Religion offers stability. Brands can emulate that stability by maintaining their quality, ensuring a consistency in venue and providing a service that can be relied on—no matter what!

The important points to focus on are the stability factors that connect the consumer with your brand. As the digital gadget market explodes, consumers are overwhelmed by instruction manuals and technical details. Navigation tools in Nokia cell phones have become the mainstay of that brand. Its users don't want to have to relearn how to use their phones with each new purchase. Since 1995, Nokia has improved and added to their phones' functionality without changing the basic system of navigation.

Like Nokia users, Apple Macintosh fans are wedded to the navigation system of their computers. PC users would struggle to find a similar fluidity in the Apple environment. Each system is familiar to its users. They know where to find their folders, how to bin useless information, how to read the clock or the hourglass—elements that are essential components of the various systems.

Another important aspect of consistency is the sales proposition—what constitutes the core product. When Japanese cars began flooding the market, extras that were optional on European and American cars were standard inclusions on theirs. So Japanese car makers gained the reputation of delivering high-performance vehicles along with the latest technical innovations, making Japan a major center of car manufacture. European cars learned from this and began to include the extras in their own cars. However, the Japanese manufacturers never changed their proposition, thereby building on the stability that consumers continue to associate with Japanese cars.

Another product, entirely unrelated, has also maintained a consistency in their sales proposition. Kinder Surprise chocolate eggs have always come with a surprise toy. It's part of the package. A Kinder Surprise would rapidly turn into a Kinder Shock if the toy was removed. If your brand contains a promise, it must never be broken. This is an essential part of building stability.

Your first step would be to identify the ten most essential components of stability currently represented by your brand. This could be your colors, your tone of voice, or even your navigation. Bear in mind that your customers have formed their own relationship with your brand, and might very well count on aspects of it that have not actually occurred to you. So it's important that you factor these needs in.

## 6. Perfect World

When mutinous crowds began to mutter and curse outside Singapore's 113 McDonald's outlets, the regional manager had to publicly plead for calm in the local press. Only the intervention of the country's baton-happy police force averted disaster.

An anti-WTO protest? Anger at American culinary hegemony? Not quite. The cause of Singaporean outrage was a three-inch plastic cat in a red ribbon and wedding dress. It was a giveaway. Part of a meal and gift deal. And the cat in question was that mouthless icon of devotion, Hello Kitty. Everyone wanted one.

The demand for the Japanese cartoon character in Asia is insatiable, and is steadily growing in the U.S. and European markets. In Hong Kong, 4.5 million McDonald's Hello Kitties sold out in a short five weeks. In Taiwan, the local Makoto Bank launched Hello Kitty credit cards, cash cards, and account books. The bank's annual revenue skyrocketed.

Hello Kitty is a savior. She draws people into her perfect world and offers stability and happiness. Nothing is more attractive than this place where chaos is removed, and you simply follow the rules—leaving it up to Hello Kitty to take care of you. Hello Kitty has twenty-five years' experience doing just that. Her devotees are free to project their every emotion onto her cute image. She has help lines, prayer sites, and private Hello Kitty counseling sessions.

Hello Kitty will also take care of your every corporal need. There are Hello Kitty tea sets, toasters, phone holders, backpacks, calendars, diaries, mouse pads, clothing, toys, motorcycles, erasers, toasters, valances, sheets, curtains, and bedspreads. Itochu Housing is selling a Hello Kitty–themed condominium to celebrate the cat's twenty-fifth birthday. Daihatsu Motor Company has produced a Hello Kitty car, complete with Kitty door locks, upholstery, and driver's console. This Japanese brand icon is a multibillion-dollar global business powerhouse. Sanrio, the company that gave birth to her, is one of the world's most successful purveyors of character-related kitsch. They have 3,500 stores in over thirty countries, and add about six hundred new products a month to the 20,000 products or so already available.

Hello Kitty is by no means the only symbol of a perfect world. Another Japanese concept has shown enormous potential. EverQuest, an online game operated by the Sony Corporation, has close to 3 million paid-up subscribers. When members were asked where they'd prefer to live—Earth or in Norath (the cyberplanet in EverQuest), 20 percent chose Norath![4] In the same way religion holds the promise of a perfect life, so clearly Norath is a very attractive alternative to our world.

EverQuest is not the only perfect place in cyberspace. As I write, there are more than 20,000 online gaming communities consisting of more than 35 million people.

Both EverQuest and Hello Kitty are extreme examples of perfect-world icons. However bizarre it may sound, they offer a strategic direction for creating a perfect-branded world. The fundamental key to success in these concepts has been the brands' ability to establish a framework of solid rules offering the consumer safety as well as freedom to reinvent themselves in a world that's more controllable, and in all likelihood more comprehensible.

They're established as a product that consumers can project their perfect-world ideas onto. Hello Kitty is able to express moods through her facial expressions. EverQuest gives its citizens space to develop their own character in the world of Norath. The player takes an active role in shaping his or her (perfect) world.

## 7. Sensory Appeal

No brand in existence can lay claim to appealing to all five senses. However, almost every religion can. Each denomination has its colors, its uniforms, its icons, and its settings.

The Church of Hagia Sophia (Holy Wisdom) was built on the highest hill in Istanbul in Turkey. Two men worked to create this astounding Byzantine structure, neither of whom had ever turned his hand to architecture before: Isidoros, an engineer and scholar, and Anthemios, an artist, experimental scientist, and mathematician. The church was built around a dome designed to lead the eye ever upward around the building, moving closer and closer to the center and finally resting on the altar.

Many religious buildings have a sense of grandeur about them, but they are also designed to impart the values driving their religion. Apart from their attempt to be a physical manifestation of the spiritual, their sacred spaces are filled with distinct smells. Burning incense in the liturgy dates back to ancient Hebrew worship, and is recorded in the Psalms. "Let my prayer be set forth in Thy sight as the incense." As this verse from Psalm 121 suggests, incense symbolizes the words of the prayer rising up to God. The Bible equates incense with visions of the Divine, most notably in the Book of Isaiah and the Revelation of St. John. The smoke itself is associated with purification and sanctification.

Incense is by no means the sole province of Christianity. The earliest use of aromatic oils and herbs has been documented in ancient China. It was also used in sacrificial religious ceremonies in ancient Egypt, Greece, and Rome. It's an integral part of all the major religions in Asia, and it is used in shrines as part of the offerings, during prayer and meditation, as well as to ward off demons and evil spirits.

Perfume starts as something magico-religious, a symbol of transformation that is confined to sacred ritual. But everywhere it develops rapidly into something far more profane as the secret spreads from the priests to the people. This transition is easy in the East, where aromatics have always played an overt part in religious ecstasy. In Tantric ritual, sandalwood oil is applied to a man's forehead, chest, underarms, navel, and groin. And a woman is similarly anointed with jasmine on her hands, patchouli at her neck, amber at her breast, musk in her groin and saffron on her feet. A wonderfully heady mix of spiritual realization and pure olfactory delight.

Religious buildings are designed to carry sound. Whether it's the organ, the choir, ringing bells, gospel music, or the sound of chanting mantras, the acoustics resonate and are an important characteristic in all houses of worship. Each religion has its own unique sound. By the same token, each has its own ceremonies, with their own symbols and rituals. The music that accompanies these rituals is a vital part of the worship—and of creating the atmosphere.

It's hard to "feel" the soul, and so each religion has devised a symbolic reference to help its adherents get in touch. The wafer on the tongue, which worshippers believe is the flesh of Christ, the vermilion tikka on the forehead marked with a firm thumb, the feel of the sacred books or the worry beads being moved around in the hands; tactile sensation is also part of the religious experience.

In direct contrast to religion, brands have to struggle to convey total sensory appeal. Perhaps this is because brands tend to narrow their focus, concentrating only on the senses that are related to the primary function of the product. In the process they ignore the value of creating a multisensory appeal.

Harley-Davidson is one of the few exceptions. The sound of their V-Twin engine has become synonymous with their brand. In 1996 the company took Yamaha and Honda to court to defend their sound. Harley's trademark attorney, Joseph Bonk, described the sound as "very fast, 'potato-potato-potato.'" Although the sound of a Hog (an affectionate term used by Harley aficionados) firing up may not be eligible for trademark protection, no doubt it is as emotional for Harley devotees as the first swelling chords of the organ that precede the beginning of mass are to the devout Catholic. Although Harley-Davidson is strategic in basing its brand recognition across several senses, most brands haven't even begun.

## 8. Rituals

Every four years a torch is lit at Olympia in Greece. It is hand carried by athletes (and celebrities) from different competing nations to wherever the Olympic Games are being held. It is an integral part of the Games' opening ceremony, continues to burn for the duration of the competition, and with much fanfare is extinguished at the close of the Games.

The ritual of the Olympic flame seemed nothing less than religious. According to the Olympic Committee, more people witnessed the ceremonial ritual of the opening of the 2000 Sydney Games in person or on television than have ever attended any religious ceremony. Throughout the two weeks of this sporting worship many other rituals unfolded. The unfurling of the flags, the music, and the award ceremonies. All following strict guidelines that have evolved over many years and are familiar to billions across the globe.

The Olympic Games are followed around the world. Whole countries stay glued to their televisions, caught up in the action of winning, losing, and overt displays of nationalism. There's drama, excitement, tragedy, and tears. The whole event comprises a series of rituals, and watching it has become a ritual in and of itself. And even after the day the flame is extinguished, symbols from the rituals live on. In any former Olympic city you'll notice signs that still direct you to the Olympic Avenue, the Olympic Square, and the Olympic Arena. Brands would pay millions for such naming placement.

Brands need rituals, although very few have managed to create them. In the heady days of rock and roll during the 1960s, Pete Townshend from The Who accidentally smashed his guitar on the ceiling of a small club. The crowd's frenzied cheers encouraged him, and so smashing his guitar became a ritual at every Who performance. It was a ritual that was subsequently adopted by another icon of the 1960s, Jimi Hendrix.

Interestingly, some of the most sophisticated brands with an abundance of rituals are relatively new, and they tend to be part of larger branded communities. Nintendo, X-Box, and PlayStation all have rituals in common. Serious gamers will tell you that they adhere to strict rituals set by the gaming community which cover everything from play patterns to cheat codes.

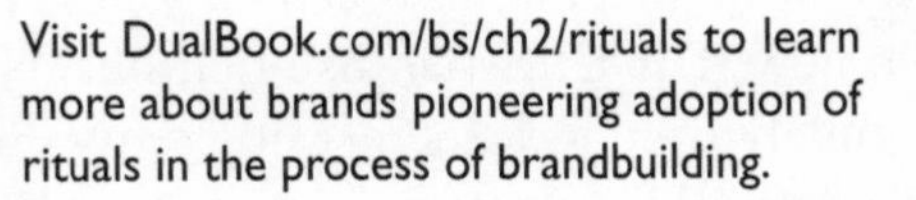


If a brand wishes to transform its traditional consumer loyalty to a community of believers, it needs to have rituals. These rituals need to embody principles of consistency, reward, and shared experience, and should be linked to a certain user situation or a particular customer need.

Consistency satisfies customer expectations and, more importantly, helps spread the word. This consistency must be meticulously executed, shared with everyone, and religiously guarded. Consistency encompasses every aspect from navigation to announcement to purpose, and must also appeal to all our senses.

Ritual must carry with it a built-in reward system. This is not necessarily some sort of financial gain; it can just as well be a pleasurable experience—pleasurable enough to repeat—again and again and again.

The most important element, however, is to ensure that the ritual is a shared one. Rituals for one don't hold much weight. One may appreciate a beautiful sunset on one's own, but the sunset is only totally splendid if there's someone to share it with. And if a whole community witnesses that sunset together, it turns into a sacred moment. The bonding in turn becomes the reward.

Over the centuries religion has managed to turn ritual into a fine art. Future brands will need to include rituals as part of their package, though this is not easily achieved. And the greatest challenge brands face will be maintaining a constancy over time in the rituals they evolve.

## 9. Symbols

The entire structure of our modern world is based on symbols. Our cell phone, video recorder, fridge, dishwasher, street signs, instruction manuals, and on and on. The list is endless, as ever more symbols enter our semiotic vocabulary. However, this phenomenon is far from new. The Christian fish (ichthus) is a symbol that has evolved. Originally used as a secret code drawn with a sandal in the sand by persecuted believers, now station wagons are adorned with holographic replicas proclaiming the drivers' Christianity to the world at large. Almost without exception, religions have sought symbols to represent and identify their faith—whether scratched in a cave, carved on a rock, or adorned with precious jewels.

Iconographic communication is on the rise. Almost all computer games function around icons. They serve two distinct purposes. Primarily they inform in a quick, simple and understandable language. They also can be used as code, recognizable only by the initiated. This forms its own code and further endorses a sense of belonging among those that adhere to it. The *BRANDchild* study revealed that 12 percent of tweens preferred written communication to verbal. An astounding 70 percent abbreviate their language on purpose when texting or chatting. As a result of this, their written language has first and foremost become expedient, accommodating icons, numbers, and nonstandard grammar.

Gangs wear their colors, motorcycle clubs their insignia, and youth concentrate on hair color, style, and fashion. So we dress, walk, and talk in a manner that shows where our affiliations lie.

Only a limited number of brands have consistently integrated symbols into their overall brand communication. Those that have, have been far from consistent. Over the past ten years Microsoft has, more than once, changed the look of some of their most well-known icons. Motorola's done the same thing. New menu settings, icons, recycle bins, books, and the like. Car brands are also guilty of discarding their symbols. Essentially they fail to capitalize on a loyalty that could serve them well across all communication channels. Symbols must reflect the brand's core values and be so distinct that it's instantly recognizable to every customer.

## 10. Mystery

What are we here for? What happens when we die? Is there life on other planets? What does God look like? Many questions have no definitive answer. But these are questions that have obsessed humankind for centuries, and continue to do so unabated.

The unknown factors in a brand have been shown to be just as inspiring as the known. It has been said that only two chemists in Coca-Cola know the formula at any one time. As the story goes, in the entire history of the company only eight people in total have known it, and only two of them are still alive. The Secret Formula refers to an ingredient called 7X, a mixture of fruit, oils and spices that gives the syrup its Coke taste. When, in 1977, the Indian government demanded the company reveal the formula, they said they'd rather forgo the gigantic Indian market than reveal its secret.

How many would have access to the good Colonel's recipe for the fried chicken that's spread its wings across the globe? The history of a brand often helps generate a mystique that attracts an audience. No one really knows if the secret KFC recipe was found when the Colonel's house was sold. The house did exist, and it did go on the market, and so when the new owner attempted to sell the "secret" recipe the story was believed.

The more mystique a brand can cultivate, the stronger the foundation it has for becoming a sought-after and admired product. Religions have been cultivating mystique since the year one. However, only a few brands have learned from the experience and made good use of this tenth rule.

---

## Highlights

As uncertainty fills our days, there's an increasing need for stability. Consumers tend to invest time and money in things they believe will survive.

For a large percentage of people, religion provides certainty in a world that's changing at an incomprehensible pace. It offers a blueprint on how to live, and provides a road map that extends way into the future, even going so far as to ensure security beyond death.

Branding continuously strives to achieve authenticity and build a relationship with consumers that will extend from cradle to grave. By its very longevity, religion automatically assumes an authentic, loyal, lifelong relationship with its adherents.

Can religion be a source of inspiration for future branding? Brands like Apple, Harley-Davidson, and Hello Kitty have already become quasi-religions—probably not religions that would pass any traditional test, but the acts of devotion they command have distinct similarities with conventional religious movements.

In order to take a brand beyond its traditional loyal base of consumers toward a bonding that resembles a religious relationship, the Ten Rules need to be followed. These are the fundamental components that underpin religion and can serve as the ultimate role model for branding:

### 1. A unique sense of belonging

Every religion fosters a binding sense of community. Within the bosom of this community belief can grow, be stimulated, and can cement relationships among the members of the congregation, creating powerful feelings of belonging.

### 2. A sense of purpose

The brand needs to reflect a clear purpose and should be represented by a visible, daring, or determined leader.

### 3. Take power from your competitors

A visible enemy gives people an opportunity to show their colors and align themselves with the team or player that they most strongly identify with—and that may often be the underdog.

### 4. Authenticity

Authenticity is an essential component of any religion's history, anecdote, and mythology—and so it should be for any successful brand.

### 5. Consistency

In an uncertain world we all yearn for stability. The important points to focus on are the stability factors that connect the consumer with your brand.

### 6. Perfect world

Brands need to establish a product that consumers can project their perfect-world ideas onto and take an active role within—further shaping this world to keep it as perfect as possible.

### 7. Sensory appeal

No brand in existence can lay claim to appealing to all five senses. However, almost every religion can. Each denomination has its colors, its uniforms, its icons, and its settings—and so should your brand.

### 8. Rituals

If a brand wishes to transform its traditional consumer loyalty to a community of believers, it needs to have rituals. Traditional celebrations—whether thought up by clever marketers or evolved over centuries—are centered around rituals.

### 9. Symbols

Iconographic communication is on the rise. All religions—and in contrast almost all computer games—function around icons.

Only a limited number of brands have consistently integrated symbols into their overall brand communication.

### 10. Mystery

Unknown factors in a brand can be just as inspiring as the known. The more mystique a brand can cultivate, the stronger foundation it has for becoming a sought-after and admired product.

---

## Action Points

- ❖ Consider all the different aspects of your brand which are similar to religion. Do you use icons, do you have a strong community, or is your brand part of ritual and tradition? Rank all these components in a diagram, and allow space for a second column.
- ❖ Now conduct exactly the same evaluation of your core competitor, and place your results in the second column. Compare your brand's performance with your competitor's.
- ❖ Few brands have succeeded in implementing all ten rules and certainly not at once. Identify the rules you wish your brand to be associated with in the future.
- ❖ Your initial evaluation should form the basis for a step-by-step analysis of the Ten Rules, and the additional components you require in order to secure a clear competitive advantage.
- ❖ Assess to what degree you intend to fulfill this objective, in terms of timing and results. This assessment should result in a detailed paper on each selected rule, with additional analysis of investment and responsibility.
- ❖ The success of your implementation will depend on your ability to take ownership of the selected rules. This will, however, get you only halfway there. The final, and yet very essential, success criterion is to ensure a strong integration between the theory you've just read about and the concept of sensory branding. The approach requires a holistic framework to succeed—and that's exactly what chapter 8 is all about.

## CHAPTER 8

# Branding: A Holistic View

**BRANDING IS EVOLVING. OVER THE NEXT** decade the dialogue will shift from better print campaigns and more catchy television commercials toward a path of reinvention. Brands will have to stand out, assert uniqueness, and establish identity as never before. Traditional advertising channels will continue to hold true, but will have to exist alongside nontraditional channels, which are mushrooming as fast as technology permits. Airwaves and cyberways are gridlocked with so many messages that it's hard to find a voice in the jam.

Short-term economic performance remains paramount. With the intense focus on ROI (return on investment) dominating every marketing move, it becomes harder to justify direct mail campaigns, when only a very low 1.61 percent of a target audience responds. And according to the Direct Marketing Association in 2003, less than 0.27 percent respond to television commercials. Ten years ago, direct mail and television ad campaigns were ten times more effective! How much longer will marketing departments be prepared to pay ten times more for advertising space which is ten times less effective than it was ten years ago?

Fifty years ago David Ogilvy, Bill Bernbach, and Stan Rapp changed the whole way we perceived advertising. They built international corporations founded on solid advertising models. Very recently we've undergone a digital revolution. We have more channels than you can poke a mouse at. We have cell phones and PDAs, the Internet and electronic games, CDs and DVDs. We have phones that take pictures, and animated images at our fingertips. We can interact with machines and people across the world in real time.

We are witnessing the emergence of the interactive consumer. By now an entire generation or two have grown up with a mouse in their hands and a computer screen as their window on the world. They respond to—if not demand—a snappier, shorter, quicker, and more direct communication.

## A Prediction

Over the next decade, sensory branding will be adopted by three categories of industry:

1. **The sensory pioneers**—automobile manufacturers and pharmaceutical companies will lead the way in sensory focus and innovation over the next decade. Trademarking components that build loyalty and avoiding expiring patents will become the main drivers.
2. **The sensory adopters**—the telecommunications and computer industries are both fighting for definition and differentiation in their commodity-driven businesses. They are most likely to look to the automobile and entertainment sectors for inspiration.
3. **The sensory followers**—a broad collection of industries including retail, FMCG, and entertainment are more likely to follow than to lead. These industries often work with smaller budgets, have less of a margin to play with, and—perhaps more importantly—are dealing with a less competitive picture than the sensory adopters.

What does it take to move into the world of sensory branding? Each and every industry has the potential to adopt a sensory branding platform; however, each one contains a set of circumstances that will influence the speed with which sensory branding can take place. Some are way ahead, others way behind.

## *Sensory Pioneers*

### Pharmaceutical Industry: Sensory Placebos

Drug companies have a limited number of years of patent protection on their products. Thereafter they're fair game for anyone to copy, and they *are* being copied. There's a steady stream of generic drugs coming out of Asia. Furthermore, tighter restrictions are being placed on pharmaceutical promotions. Marketing departments may find that sensory branding will provide a solid base from which to create a platform for the next generation of points of differentiation.

The pharmaceutical industry has reached a point where the protected patented period is insufficient to earn back the very high costs of research and development. As such, they need strong loyalty-building features to bond with the consumer in order to extend the shelf life of their products.

There's a distinct relationship between branding and placebos. In each case it is believed that the brand adds value to a product or service. This may or may not be true. In the same way that consumers believe a drink tastes better in a bottle than a can, so these factors come into play in the pharmaceutical industry.

Leveraging consumer loyalty by means of the tactile feeling of the product, its packaging, colors, and package design as well as the distinct sound, aroma, and flavor can provide the pharmaceutical company with a whole new geography for bonding with their consumers. Regulations in some countries challenge traditional trademarks on the shape and color of medication, but no trademark on smell or taste for such products has so far been rejected. This opens up a welcome vista for companies that can count on a lifetime trademark instead of a patent with a set expiration date.

### Automobile Industry: The Controlled Sensory Experiment

Sometimes it's hard for leaders of the pack to stay ahead. The car industry is now moving into the last phase of innovative sensory branding. They're working on new sounds for seat adjustments, gearboxes, rails, indicators, hazard warnings, horns, and electric windows as well as designing a low-noise, branded-sound car cabin.

Every possible component on the car that represents a sensory touch point is being scrutinized, evaluated, and branded. Soon every car brand will have its own branded smell, a branded tactile feeling, as well as sound. It won't be long before each component will be trademarked and exclusive to the model and brand. Then the manufacturer can take its trademarked components to market and extend them into a new universe of product merchandising. Porsche already has a diverse range on the market. You can buy anything from Porsche umbrellas to Porsche glasses. The sensory touch points become the primary point of contact and connection—perhaps explaining why people who are Porsche aficionados are prepared to pay 40 percent more for a Porsche laptop than any other brand.

Visit *www.DualBook.com/bs/ch7/pioneers* to learn more about which industries in the future will be the leaders in terms of adoption of sensory branding.

## *Sensory Adopters*

### Telecommunications: More than Technology

The global struggle for telecommunication dominance is reminiscent of the car manufacturers' struggle in the mid-twentieth century. Again, Asia is taking a back seat to European and U.S. companies in terms of innovation—this time in the cell phone market. And again, the Asian manufacturers are poised to bring high-standard multisensory perspectives to the product.

Every aspect of the phone, from the tactile qualities, the design and display, the branded sounds generated when using the phones, to even the smell of the product, will be evaluated, enhanced, and improved over the next few years. As further integration across personal computers, PDAs, and cell phones takes place, so the sensory touch points will move across platforms bonding the products together and leveraging the glue of branded sensory touch points. Likewise, as technology presents new innovations, these can be integrated immediately into the sensory world of the brand. An example is Immersion, whose technology is designed to allow you to "touch" someone over the phone. According to BBC online, "The company has been talking to mobile manufacturers to build in touch into future phones."[1]

## Computers

Computers have adopted the term "sound quality" from the automobile industry. This marks only the beginning of the race to gain the competitive advantage on every aspect but the size of the microprocessor.

Apple and Bang & Olufsen are providing the inspiration for an industry that has only recently become preoccupied with style and design. They're also focusing on sound. Next to tackle are the tactile elements, followed finally by the smell of the equipment. In the same way it has become standard to leverage smell in cars, so computer brands will invent their very own versions of a new-computer smell.

In contrast to so many other industries, technological innovations are built in to the product. Soon computers will be manufactured with a capacity to handle sensory channels. Since 400 million people around the world turn on a computer each day, focus is on the mouse to potentially house the multisensory "brain." Sony Corporation is working on it.[2] A team of experts, including a psychologist, are developing a mouse that will "feel" like what it's pointing to on the screen. The mouse could be installed on any computer with Windows, and would deliver images, text, and animation directly to the fingertips. Although this technology is being designed primarily for visually impaired people, the potential for other applications—namely sensory branding—is enormous.

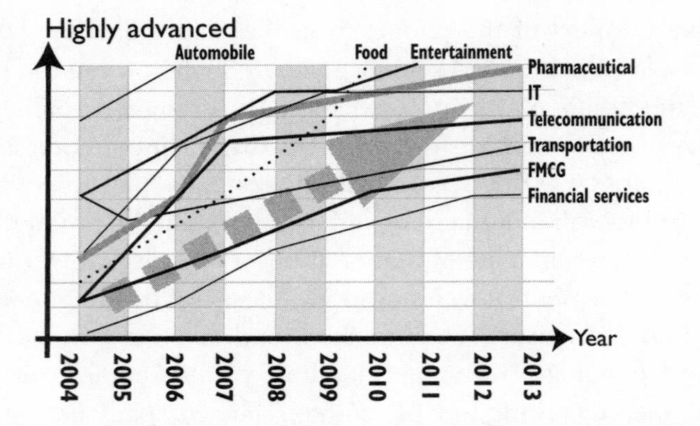


**FIGURE 8.1** Within five years it is estimated that 35 percent of all Fortune 100 brands will have established a sensory branding strategy.

LCD screen technology has taken a step further, allowing the user to feel and push virtual buttons. Sony Corporation is producing such a panel as a breakthrough, next-generation device that could potentially replace the conventional keyboard with a system of touch panels.

These innovations are still on the drawing board, but they clearly show the direction computers are heading. By incorporating touch into their product they are allowing users to "feel" brands and giving the screen a tactile quality.

## *Sensory Followers*

### Sensory Food Design

However we feel about tampering with the genetics of what we eat, the concept of food design will dominate the next decade. Taste is important, but the evidence from the *BRAND sense* study shows that smell and appearance are ranked together on the scale of importance.

The food industry is unlikely to leave things be. They will persist in designing the smell of the product and the sound of the packaging. They will also seek to control the sound the food makes when being eaten. They will tamper with color and the flavor, creating new levels of sensory preference. Tweens will like their ketchup green and their Sprite turquoise.

In our contemporary urban society we're more familiar with picking apples from supermarket shelves than from trees. Not many more than a handful of adults, let alone kids, would be able to identify an apple leaf. Although most consumers like the odor of what they believe is real leather, what they were introduced to a generation ago was a fake leather smell that they now take for the real thing. We now perceive the altered world as more authentic than the original. Technology has enabled companies like Nestle, Coca-Cola, and Carlsberg to add aroma to packaging on the supermarket shelf. The concept of authenticity will decide how far industry can go before running into a consumer backlash.

## FMCG: More than Design

The FMCG category is broad. It includes everything from toilet brushes to pens. Some industries within it will move along the sensory path more easily than others. Through the work of designers like Terence Conran and Philippe Starck, everyday items have become increasingly sophisticated visually. The next step will include differentiation of scent and sound from one brand to another. To survive in this new sensory-defined landscape companies will have to take their cue from more advanced industries to keep a lead in their own.

## Travel and Hospitality: The Sensory Glory Has Faded Away

Up until the close of the twentieth century, the hospitality industry was one of the most innovative leaders in the sensory branding department. Financial crises, SARS, terrorism, and the consequent wariness about travel are factors that have contributed to stalling their lead.

Only a few hotel chains are maintaining their sense focus. The Ritz Carlton is one. Their lion logo is visible on door handles, cake towers, soap, and slippers. However, despite the general loss of focus, the chains will remain innovators, with the Asian groups—particularly the Singaporean ones—at the forefront.

The travel industry has taken a hit all around. Budget airlines have forced regular airlines to cut their branding budgets to the bare bones. Despite these crises, a few airline companies—Cathay Pacific, Singapore Airlines—have managed to keep their sensory touch points alive. Interestingly, it's these two companies that have shown the clearest signs of recovery, placing them in an exclusive league of profitable airlines.

## Financial Institutions: Rediscovering the Personal Aspect

As banks merge and grow, so the individual becomes increasingly insignificant, creating an ever widening gap between institution and individual. Only restoring a human touch can reestablish the bond, and sensory branding will create one of the connections.

The banking retail environment has become increasingly automated. Costs have been passed on to consumers who prefer dealing face to face rather than conducting their transactions through ATMs, the telephone, automated websites, and voice machines. In stark contrast, many other retail businesses have chosen the opposite strategy, creating cozier, friendlier, lighter, more welcoming, and more branded environments.

Financial institutions now deal in commodity products. The days of a friendly manager with a reassuring smile and a warm handshake are fast disappearing. Customer loyalty in the current banking environment is as stable as today's share price. Sensory branding may be the banks' only route back to a human-centered environment.

## Retail: From Nonbranded to Branded Sensory Environments

The sensory progress of retail has been steady over the past decade. First music was introduced in some stores, then environmental designers altered layout and decor, and now they're making use of aromas. However, all this sensory progress is nonbranded. Few chains are developing their own branded sound and tactile-designed bags and packaging. The stimulus still tends to be added randomly, and revolves more around generating traffic than generating loyalty. Over the next couple of years we should see this trend convert from nonbranded activities to branded. Each branded stimulus and branded enhancement will support the differentiation of each retail chain.

Technology will also move retailers in the right sensory direction. Sonic branding is about to enter the next generation of sensory branding. Sonic logos will be incorporated into packaging that will sound branded tunes when opened. Nonbranded sonics are already in operation at the Hong Kong airport. Just as the escalators there tell you when it's time to step off, so a voice will let you know when the next checkout will be available in the supermarket.

Technology has the means to create sonic showers. A sonic shower is a narrowly delineated space where you can hear sound. The moment you step out of its invisible field, the sound is no longer audible.

## Fashion: A Total Sensory Experience

In 2002 Prada revolutionized dressing rooms in its Soho store in New York City, where it installed smart closets. Smart closets scan the individual electronic chip–based clothing tags and send the garment information to an interactive touch screen in the cubicle. The customer can then use the screen to select other sizes, colors, or fabrics. The screen also displays video footage of the garment being worn on the Prada catwalk.

Retail and fashion have merged to form an entertainment experience, leveraging technology that communicates via more senses. The microchips are able to identify an "anti-color clash," which will tell shoppers if a new garment will match their existing clothes. Once they buy an item, a chip would be able to tell them the way to take care of it.

The fashion industry is fast catching up to the perfume industry in the sensory stakes.

## Entertainment: Sensory Schizophrenia

Increasingly more merchandising programs travel on the back of a movie. Many a movie has a ride in a theme park. The entertainment industry is doing well in the sensory branding department. However, the challenge for longevity remains. On average one movie has a financial lifespan of six months. When the box office draw declines, the ride or game will lose its relevance, making it hard to justify an Indiana Jones or Harry Potter ride in Disneyland or Warner Brothers World permanently.

The sensory branding synergies between movies, cinemas, merchandising, theme parks, and events is often questionable. There are more than three thousand items of merchandise under the Harry Potter umbrella. They often share little except the fact that they're made in China and have the Harry Potter logo. Harry never invented his own smell. Neither was the brand characterized by a special sound, touch, or taste. The merchandising appeals only to the eye. It has no sensory links to the movies, the rides, or the book. It's just another piece of merchandising, which probably won't survive beyond the franchise's lifespan.

## Gaming

Computer games are fearlessly venturing on into the sensory universe, leveraging technology as the tool of implementation. They seek to simulate the real world. Tetris, the 3-D game, will soon come with surround sound and tactile stimulation. There are more than 100 million gamers out there, providing all the motivation inventors and technology companies need to bring on as many senses as possible.

Over the next few years the computer gaming industry will push mass sensory communication even further by introducing a variety of mice and joysticks to a world where 30 percent of those who play computer games do so several times a week.

Real tactile experiences are already a reality. The Immersion Corporation has released TouchWare Gaming, which they promote as a "Touch sense technology [which] can transform any game into a multi-sensory experience by engaging your sense of touch." TouchWare Gaming is already for sale to consumers. With it you can "feel your light saber hum" and "your shotgun blast and reload." You will also know if "your missile locks on a target" or if your car is "driving over cobblestones."[3]

The Nostromo n30 mouse looks like any other rollerball mouse with a stylish black paint job. However, what you see is not what you get, because this mouse is embedded with TouchWare technology. In sync with the visuals on the screen, the mouse can whirr through a palette of vibrations, which are picked up by the fingertips. Sony PlayStation's game controller offers a different type of feedback—"rumble"—and this allows the gamer to feel every bump, impact, and crash in whatever game they're playing. The SideWinder Force Feedback 2 joystick from Microsoft supports force feedback—the sensation that users feel in their hands as they play certain games.

## *Sensory Excellence*

### The World's Top Sensory Brands

Based on input from focus groups across the world, we analyzed the world's top brands from a sensory excellence point of view. Among the two hundred most valuable brands according to Interbrand, it became clear that a very limited number of brands today leverage their sensory potential. In fact, less than 10 percent of these brands demonstrate a sensory branding platform, although this is expected to increase to 35 percent within the next five years.

We used the following criteria to assess two hundred most valuable brands:

- Is the brand leveraging available sensory touch points?
- Is there a strong, consistent synergy across each of the touch points?
- To what degree does the brand reflect an innovative sensory mind-set that sets it apart from its competitors?

| RANK | BRAND | SENSORY LEVERAGE (in percent) |
|---|---|---|
| 1 | Singapore Airlines | 96.3 |
| 2 | Apple | 91.3 |
| 3 | Disney | 87.6 |
| 4 | Mercedes-Benz | 78.8 |
| 5 | Marlboro | 75.0 |
| 6 | Tiffany | 73.8 |
| 7 | Louis Vuitton | 72.5 |
| 8 | Bang & Olufsen | 71.3 |
| 9 | Nokia | 70.0 |
| 10 | Harley-Davidson | 68.8 |
| 11 | Nike | 67.5 |
| 12 | Absolut Vodka | 65.0 |
| 13 | Coca-Cola | 63.8 |
| 14 | Gillette | 62.5 |
| 15 | Pepsi | 61.3 |
| 16 | Starbucks | 60.0 |
| 17 | Prada | 58.8 |
| 18 | Caterpillar | 57.5 |
| 19 | Guinness | 56.3 |
| 20 | Rolls-Royce | 55.0 |

**FIGURE 8.2** An extensive evaluation of the world's 200 most valuable brands reveals members of this exclusive club.

- To what extent does the consumer associate these sensory signals with this particular brand—and how authentic do they perceive these signals to be?
- How distinct and integrated are these signals for the consumer?

Fourteen of the top twenty global brands do not feature on the chart of leading sensory brands. There are three main reasons for this discrepancy:

1. There's too big a gap between the marketing department and the research and development department. Brands like Singa-

pore Airlines, Disney, and Apple have a very strong integration between these divisions, and they all work together.
2. A sensory appeal has not yet been defined as essential among these companies, nor has it examined the major potential that sensory branding can unleash.
3. Some industry categories are more naturally predisposed to such an approach, and even though it has been given due consideration, it may have been deemed unsuited to further pursuit.

The most interesting fact to emerge is that the majority of top twenty brands that do leverage a multisensory platform have even more potential than what we've seen to date. The Marlboro cowboy's rugged image is an excellent sensory vehicle to spread across product lines. Louis Vuitton's steady growth of merchandise gives it carte blanche to ensure a four-, if not five-, sense appeal value. Nokia's steady growth of digital channels represents many sensory opportunities to leverage the company's icon, sound, and navigation features. Gillette needs to focus on their inconsistencies in tactile and aroma-based signals, and Starbucks still has a way to go to optimize sensory appeal in their many cafés, where their lines of merchandise tend to be neglected—and where it today is a fact that their customers, according to the *BRAND sense* study, don't relate a distinct taste to Starbucks at all.

## The Untapped Potential: The Bottom Twenty Sensory Brands

But the true untapped sensory potential can be found in Figure 8.3, listing the top twenty brands with the largest untapped sensory potential.

## But This Is Just the Very Beginning . . .

Even if you can tick every sensory box on the checklist for your brand, claiming that every sensory aspect of your brand has been fulfilled is far from the end of your sensory story. However, it creates

| RANK | BRAND | SENSORY LEVERAGE (in percent) |
|---|---|---|
| 1 | Ikea | 23.8 |
| 2 | Motorola | 25.0 |
| 3 | Virgin | 26.3 |
| 4 | KFC | 28.8 |
| 5 | Adidas | 31.3 |
| 6 | Sony | 31.3 |
| 7 | Burger King | 31.3 |
| 8 | McDonald's | 32.5 |
| 9 | Kleenex | 32.5 |
| 10 | Microsoft | 33.8 |
| 11 | Philips | 33.8 |
| 12 | Barbie | 33.8 |
| 13 | Nescafé | 35.0 |
| 14 | Nintendo | 36.3 |
| 15 | Kodak | 40.0 |
| 16 | AOL | 41.3 |
| 17 | Wrigley | 42.5 |
| 18 | Colgate | 43.8 |
| 19 | IBM | 45.0 |
| 20 | Ford | 46.3 |

**FIGURE 8.3** Top 20 brands with the most untapped sensory potential. Many top brands have, so far, failed to capitalize on their sensory potential.

the perfect basis for a true upgrade of your brand, securing a new level of differentiation almost impossible for any brand to imitate it on the rise.

## The Holistic Sales Proposition (HSP)

Currently brand manufacturers own their brands. This is changing. In future brands will increasingly be owned by the consumer. The first signs of this shift appeared in the late 1990s. I documented this phenomenon in *BRANDchild* and named it MSP—Me Selling Proposition.

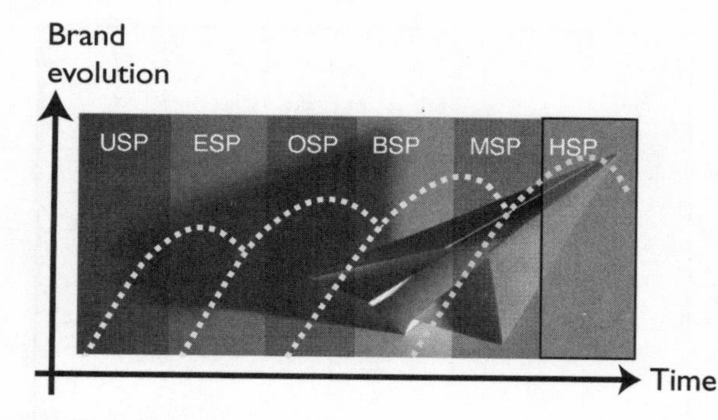


**FIGURE 8.4** Since the very beginnings, the branding concept has evolved from a USP (Unique Selling Proposition) strategy to an MSP (Me Selling Proposition) strategy. The HSP (Holistic Selling Proposition) is likely to dominate the next decade of brands—brands that are heavily inspired by sensory and religious trends.

As I said in chapter 1, the history of branding has passed through numerous shifts from USP (Unique Selling Proposition), through ESP (Emotional Selling Proposition), OSP (Organization Selling Proposition), and BSP (Brand Selling Proposition) to today's MSP ("Me" Selling Proposition).

## The World of Holistic Branding

There's every indication that branding will move beyond the MSP into an even more sophisticated realm. I call this realm the HSP, the Holistic Selling Proposition. HSP brands are those that not only anchor themselves in tradition but also adopt some of the characteristics of religions to leverage the concept of sensory branding as a holistic way of spreading the news. Holistic brands are smashable. They have their own identity, which is expressed in its every message, shape, symbol, ritual, and tradition.

NASA named its first space shuttle *The Enterprise*. This was not based on any corporate name-finding strategy, but was the result of 400,000 requests from *Star Trek* fans all around the world. *Star Trek* was more than a television show, it grew into a holistic, all-enveloping brand complete with its own language, characters, sounds, and design. Star Trek is a brand with a religious following. Few brands

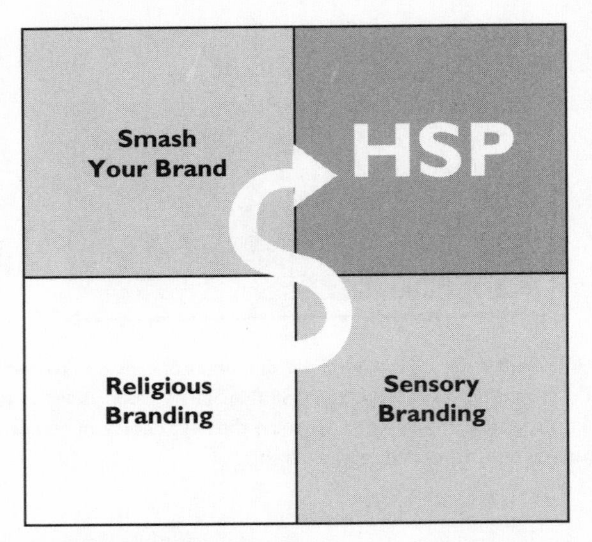


**FIGURE 8.5** The Holistic Sales Proposition. Transforming your brand into an HSP position requires a systematic approach. Any HSP brand is based on one or several of the ten brand rules outlined in chapter 7. In addition, an HSP brand should be truly smashable and leveraged across all accessible sensory touch points.

have succeeded in turning customers into evangelists. Not many companies can lay claim to this, although customers' belief in the brand is the most essential building block.

## Characteristics of an HSP Brand

- A true HSP brand is not logo-centric. Its power is embedded in every aspect of the brand. Its message, sound, smell, and touch let you know what it is.
- An HSP brand leverages every possible channel to communicate its philosophy. An HSP brand is as visible in online communities as it is in the national press.
- Consumers claim ownership, rather than the company. The survival of the HSP brand is important to its consumers. If the brand were to experience difficulty of any sort, the consumers would rush to its rescue.

- The HSP brand is enhanced by the attributes bestowed upon it by its consumers who develop rituals and traditions around its usage.
- The brand has its clear and well-defined enemies, clear and well-defined leaders, and clear and well-defined followers.
- An HSP brand has a distinct history, with ups and downs, historic moments, and major celebrations. It can spark a conversation that would still be of interest ten minutes later.
- Customers would consider wearing an HSP brand as body decoration.

The HSP philosophy recognizes the interconnectedness of all the elements that go to make up the brand. Every communication it sends out through every channel is related to the brand's core philosophy. Every component is a vital piece that creates the full holistic picture.

To form a true HSP brand presents a rigorous challenge. Few companies can consider their brand as belonging in the HSP category, although in principle every brand can get there. Brands, like humans, evolve over time. The story is only beginning. Once on the path, the journey toward reaching a holistic brand promises to be interesting.

---

### Highlights

Less than 10 percent of the world's top brands demonstrate a sensory branding platform, although this is expected to increase to 35 percent within the next five years. The development will happen within the following three categories of industry.

1. **The sensory pioneers**—the automobile and pharmaceutical industries will lead the way in sensory focus and innovation in the next decade. Trademarking components that build loyalty and avoiding expiring patents will become the main drivers.

2. **The sensory adopters**—the telecommunications and computer industries are both fighting for definition and differentiation in their commodity-driven businesses. They are

most likely to look to the automobile and entertainment sectors for inspiration.

3. **The sensory followers**—this, a broad collection of industries, including retail, FMCG, and entertainment, is more likely to follow than to lead. These industries often work with smaller budgets, have less of a margin to play with, and perhaps more importantly, are dealing with a less competitive picture than the sensory adopters.

The future of sensory brands will be evaluated on the following criteria:

- Is the brand leveraging available sensory touch points?
- Is there a strong, consistent synergy across each of the touch points?
- To what degree does the brand reflect an innovative sensory mind-set that sets it apart from its competitors?
- To what extent does the consumer associate these sensory signals with this particular brand—and how authentic do they perceive these signals to be?
- How distinct and integrated are these signals for the consumer?

There's every indication that branding will move into an even more sophisticated realm. I call this realm the HSP—the Holistic Selling Proposition. HSP brands are those that not only anchor themselves in tradition but also adopt religious characteristics and leverage the concept of sensory branding as a holistic way of spreading the news. Holistic brands are smashable. Each has its own identity which is expressed in its every message, shape, symbol, ritual, and tradition.

---

### Action Points

You now have a solid framework for a sensory branding strategy. Check how your strategy compares to others in your industry, as well as how it measures up to the industry you benchmarked. Is your strategy solid?

- ❖ Have you applied the ten rules of sensory branding to leverage sensory potential in a way that reinforces your brand values, brand alliances, and brand extensions?
- ❖ Are you staying true to the brand alliance and extension constitutions?
- ❖ Can I smash each of the many new sensory components you've identified?

To become a true HSP brand takes time and patience. Check out DualBook.com, which will update you monthly with the latest movements within the world of Sensory Branding.

# Notes

## 2. Some Companies Are Doing It Right

1. Al Ries and Laura Ries (2002), *The Fall of Advertising and the Rise of PR,* HarperCollins, New York.
2. TV Turnoff Network, www.tvturnoff.org.
3. *Fortune* magazine, June 28, 2004.
4. Ibid.
5. Ries and Ries, *The Fall of Advertising and the Rise of PR.*
6. David Shenk (1998), *Data Smog: Surviving the Information Glut,* HarperCollins, New York.
7. Newspaper Advertising Bureau.
8. H. A. Roth (1988), "Psychological Relationships Between Perceived Sweetness and Color in Lemon-and-Lime Flavored Drinks," *Journal of Food Science,* 53:1116–1119.
9. C. N. DuBose (1980), "Effects of Colorants and Flavorants on Identification, Perceived Flavor Intensity, and Hedonic Quality of Fruit-Flavored Beverages and Cake," *Journal of Food Science,* 45:1393–1399, 1415.
10. Quoted in Diane Ackerman (1990), *A Natural History of the Senses,* Vintage Books, New York, p. 191.
11. Ibid., p. 5.
12. Lyall Watson (2000), *Jacobson's Organ: And the Remarkable Nature of Smell,* W.W. Norton & Company, New York, p. 7.
13. Ibid., p. 88.
14. Ibid., p. 90.
15. Ibid., p. 136.
16. Boyd Gibbons (1986), "The Intimate Sense of Smell," *National Geographic* (Sept.), p. 324.
17. Ashley Montagu (1986), *Touching: The Human Significance of the Skin,* 3rd ed. Harper & Row, New York, p. 238.
18. www.innerbody.com/text/nerv/G.html.

## NOTES

### 3. Smash Your Brand


1. John Hagel and Marc Singer (1999), *Net Worth,* Harvard Business School Press, Boston.
2. BRAND sense study, 2003.
3. www.benetton.com.
4. www.quickstart.clari.net/qs_se/webnews/wed/bz/Bpa-heinz.RaYS_DSA.html.
5. www.press.nokia.com/PR/199810/778408_5.html.
6. Ibid.
7. www.absolut.com.
8. www.fredericksburg.com/News/FLS/2002/102002/10032002/747192.


### 4. From 2-D to 5-D Branding


1. Discovery Communications Inc., 2000.
2. BRANDchild study carried out by Millward Brown, 2002.
3. Ronald E. Millman (1985), "The Influence of Background Music on the Behaviour of Restaurant Patrons," *Journal of Consumer Research,* vol. 13.
4. Judy I. Alpert and Mark I. Alpert (1988), "Background Music as an Influence in Consumer Mood and Advertising Responses," in Thomas K. Scrull (ed.), *Advances in Consumer Research,* vol. 16, pp. 485–491.
5. Kevin Ferguson, "Coin-free slot jackpots? Unclinkable!" http://www.reviewjournal.com/lvrj_home/2000/Aug-28-Mon-2000/business/14239785.html.
6. Richard E. Peck (2001), "Bill Gates' Bite of the Big Apple," www.ltn-archive.hotresponse.com/december01/.
7. Diane M. Szaflarski, "How We See: The First Steps of Human Vision," www.accessexcellence.org/AE/AEC/CC/vision_background.html.
8. Sarah Ellison and Erin White (2000), "Sensory Marketeers Say the Way to Reach Shoppers Is by the Nose," *Financial Express,* Nov. 27.
9. www.hersheypa.com/index.html.
10. Ken Leach, *Perfume Presentation: 100 Years of Artistry,* quoted on www.wpbs.com.
11. Warrem and Warrenburg (1993), "Effects of Smell on Emotions," *Journal of Experimental Psychology,* 113 (4):394–409.
12. Amanda Gardner (2003), "Odors Conjure Up Awful 9/11 Memories," www.healthfinder.gov/news/newsstory.asp?docID=513682.
13. H. A. Roth (1988), "Psychological Relationships Between Perceived Sweetness and Color in Lemon-and-Lime Flavored Drinks," *Journal of Food Science,* 53:1116–1119.
14. Christopher Koch and Eric C. Koch (2003), "Preconceptions of Taste Based on Color," *Journal of Psychology* (May), pp. 233–242.
15. Trygg Engen (1982), *Perception of Odors,* Academic Press, New York.

16. Stephan J. Jellinek (2003), "The Underestimated Power of Implicit Fragrance Research," and Peter Aarts (2003), "Fragrances with Real Impact," papers presented at Fragrance Research Conference, Lausanne, March 16–18.
17. www.theecologist.org/archive_article.html?article+342&category=33.

### 5. Stimulate, Enhance, and Bond

1. Katie Weisman (2003), "Brands Turn Onto Senses," *International Herald Tribune, online* www.iht.com/articles/120122.html, Dec. 4.
2. Ibid.
3. Carlson School of Management, "Creating Value Through Strategic Alliances," www.csom.umn.edu/Page1623.aspx?print=True.

### 6. Measuring Senses

1. Paul Dyson, Andy Farr, and Nigel Hollis (1996), "Understanding, Measuring and Using Brand Equity," *Journal of Advertising Research* (Sept./Oct.).
2. Larry D. Compeau, Dhruv Grewal, and Kent B. Monroe (1998), "The Role of Prior Affect and Sensory Cues on Consumers' Affective and Cognitive Responses and Overall Perceptions of Quality," *Journal of Business Research,* 42:295–308.
3. "Making the Most of Your Brands" (2002), Page, Admap (Nov.).
4. P. R. Kleinginna and A. M. Kleinginna (1981), "A Categorized List of Emotion Definitions with Suggestions for Consensual Definition," *Motivation and Emotion,* 5: 345–379.
5. "The Pleasure Seekers," *The New Scientist,* Oct. 11, 2003.
6. Hollis (1995), "Like It or Not, Liking Is Not Enough," *Journal of Advertising Research* (Sept./Oct.).
7. Alan Branthwaite and Rosi Ware (1997), "The Role of Music in Advertising," *Admap* (July/Aug.), p. 44.

### 7. Brand Religion

1. David Lewis and Darren Bridger (2001), *The Soul of the New Consumer: Authenticity—What We Buy and Why in the New Economy,* Nicholas Brealey Publishing, London, p. 13.
2. http://www.isn.ne.jp/~suzutayu/Kitty/KittyPray-e.html.
3. Matthew W. Ragas and Bolivar J. Bueno (2002), *The Power of Cult Branding: How 9 Magnetic Brands Turned Customers into Loyal Followers (and Yours Can Too!),* Prima Venture, New York, p. 28.
4. Sean Dodson (2002), "The World Within," *Sydney Morning Herald,* Icon, May 23.

## NOTES

### 8. A Holistic View

1. Alfred Hermida (2003), "Mobiles Get a Sense of Touch," www.news.bbc.co.uk/1/hi/technology/2677813.stm, Jan. 21.
2. www.vtplayer.free.fr.
3. Immersion Corp., "Feel the Game with TouchWare Gaming," www.immersion.com/gaming.

# The BRAND sense Research

In 2003 Martin Lindstrom came to Millward Brown, a leading and innovative global market research agency specializing in helping companies maximize their brand equity and brand performance. He had an unusual request: "Help me prove that the sensory experience of brands plays a key role in creating brand loyalty." While our clients around the world have approached us with many questions related to the effectiveness of their brand-building and marketing activities, this one was unique. We experience the world through our senses, so intuitively it seemed obvious that brands could create a stronger emotional bond by differentiating their sensory appeal. The question was, could we prove it?

To do so we designed a two-stage research program that spanned the globe, involving hundreds of researchers and talking to thousands of people.

## Stage One: Understanding the Role of the Senses

When tackling a new and unique project like this, it is critical to understand the mental "landscape" in which brands exist. Qualitative research, where a trained moderator explores ideas and brand associations using projective techniques with small groups of people, is invaluable, providing insights and guiding the way to a more quantitative measurement.

We conducted focus groups in thirteen countries: Chile, Denmark, Holland, India, Japan, Mexico, Poland, Spain, South Africa, Sweden, Thailand, the United Kingdom, and the United States. In each country, we talked to men and women aged twenty-five to forty. The exploration focused on ten global brands: Coke, Mercedes-Benz, Dove, Ford, Gillette, Vodafone/Disney, Levi's, Sony, Nike, and McDonald's. Five additional local brands (which varied by market) were also included.

The findings from the research gave us a good understanding of the role of the senses in creating brand loyalty, and confirmed that the brands with sensory depth were particularly strong, with clearly defined, globally understood, and distinctive brand identities and with relevant and aspirational brand values. In some respects at least, these brands had deliberately built their sensory values, and were now benefiting from owning such associations.

## Stage Two: Quantifying the Influence of the Senses

In many ways this was the most challenging phase of the research. We now wanted to prove that the memory of a brand's sensory associations led to a higher intent to buy the brand.

To do so, we created and tested a unique online questionnaire, as described in chapter 6. In conjunction with our partner, Lightspeed Online Research, we interviewed over two thousand people in the United States, the UK, and Japan. They provided feedback on their sensory associations, imagery, purchase intent, and much more, for eighteen brands.

We then used a statistical methodology called Structural Equation Modeling to test hypotheses (from stage one) on how the senses might affect brand loyalty. Several models based on the varying hypotheses of how the variables interrelate were developed for the same set of data. Each model was then evaluated by a combination of diagnostics associated with its individual paths as well as an overall Goodness of Fit Index. The model with the best overall fit and intuitively sensible paths was chosen as the best representation of reality. Again, the results are reported in chapter 6.

*Robert D. Meyers*
*Group CEO, Millward Brown*

# A Few Words from the Researcher

We all know the trends. Number of brands up. Price competition up. Media options up. The barriers to marketing success are becoming ever higher. Marketers today face a tough job trying to keep their brands healthy and profitable. Millward Brown's mission is to help them do so by providing insights into how to build and maintain brands in this increasingly complex world, which is why we were eager to develop the research findings for this book.

Today's complex marketing world requires us to understand the impact of all the different influences on a purchase decision.

Take the case of digital cameras. Job number one is to make sure that your brand is included in the consideration set when people first start thinking about a purchase. That means sowing the seeds early, through traditional advertising, viral marketing, and publicity. No one, however, buys a camera without regard for price. Send the wrong signals and you can easily get ruled out simply because people think you are too expensive.

Good job, you are now on the shopping list. Now it is down to features and price, right? Wrong. Very few brands get rejected on the grounds of performance or price. Almost every brand of camera offers a wide range of features and price points so that most people can find something that meets their needs and budget. What, then, determines the purchase decision? For many it is how the camera looks, feels, and sounds. Does it feel good to use? Does it look cool or purposeful? Does it make the right noises? Based on their experience of using film cameras, people expect to hear a click and a whirr when taking a photograph. Not hearing that sound made people uncomfortable with some of the early digital cameras. The latest digitals use sounds reminiscent of film cameras to signal that a photograph has indeed been taken. It is the little things that swing the purchase decision.

It is the very lack of anything other than a visual experience that prevents many people from buying online. Even the most aggressive proponents of Internet shopping limit the potential size of the market because sensory perception is so very important. In the case of digital cameras only one in four recent purchasers in the United States claim to have bought their camera online. In the case of automotive brands, people use the Internet to research facts, options, and prices, but they all visit the showroom to make the final decision. It's the responsiveness of the car to the controls, the seat comfort, even the smell, that closes the deal. Buying a new car is both a serious decision and a sensuous experience. To buy a car based on sight alone would leave most people dissatisfied and very worried that they had made a mistake. Whatever the product or service, most people will always want to experience touch, smell, sound, and taste, as well as visual appeal, before they buy.

In the modern day and age marketers have almost forgotten the power of the senses, favoring instead the cool rationality of product specifications and the cut and thrust of price discounts. That's why the senses offer a potent means of communication, helping marketers to find new ways to differentiate their brands and strike an emotional chord. The senses are such a fundamental part of being human that they are inescapable. They influence us every second of the day. Marketers who recognize the power of the senses will find a new means to build a long-lasting bond with their consumer. Not one based on discounts and loyalty programs but one based on enjoyment and appreciation.

Martin Lindstrom has explored the sensory arena with the same enthusiasm that he has displayed when tackling the Internet or the lives of today's tweens. Our research was designed to help him illustrate the impact of the senses and to demonstrate how the senses impact brand choice and loyalty. I believe the combination has made this book a great read which will help you see the marketing world with new eyes.

*Nigel Hollis*

# Acknowledgments

I am indebted to Nigel Hollis for providing chapter 6. Nigel is the global strategic planning director for Millward Brown. He has extensive experience in market research, and his particular expertise includes advertising pre-testing, brand equity research, online research, and how marketing communications can build and maintain brands. His career at Millward Brown has spanned the Atlantic; he has worked for major Fortune 500 clients, covering packaged goods, automotive, alcoholic beverages, financial services, IT, and travel categories.

It would be impossible to mention everyone else who contributed to the ideas and knowledge I've drawn on for this book. There are close to six hundred researchers who have assimilated data, provided insight, and worked hard to make *BRAND sense* a truly global project.

First and foremost, Lynne Segal (Australia). It's the third book we've worked on together, and, in short, if Lynne quit editing, I'd quit writing. Also I owe a sincere thanks to my superb agent James Levine; my wonderful editor Fred Hills and the impressive team at Free Press, including Suzanne Donahue, Carisa Hays, and Michele Jacob—all from the United States.

Once again the Millward Brown organization has been a pleasure to work with. Without their support for this global project, it would have been almost impossible to execute. They have verified my assumptions, and taken the research where no other marketer has gone before—into a tactile world full of smells and tastes. Nigel Hollis (U.S.) managed the research project with skill, professionalism, and perpetual optimism. I would also like to thank Andreas Sperling (Singapore) for his abundant energy and perfect (almost German) management of this project. Bob

Meyers (U.S.), Eileen Campbell (U.S.), Andrea Bielli (Italy) and Sue Gardiner (UK) have given enormous support behind the scenes. Thanks to Jean McDougall (UK) for her belief in all my marketing ideas, Andreas Grotholt (Germany) for his ongoing feedback, and Andreas Gonzales (Australia) for his generosity from the very first day I knocked on his door.

A very special thanks to Ulrik With Andersen and Birgitte Rode, both from Audio Management, who helped me with excellent input on sound and support setting up the *BRAND sense* show. They are a truly remarkable team.

Several people around the world have helped me capture the essense of sensory branding. These include Karen Elstein (UK), Andres Lopez and Claudia Jauregui (Mexico), Mauricio Yuraszeck, Marco Zunino, and Maria Cristina Moya (Chile), Chaniya Nakalugshana and Tanes Chalermvongsavej (Thailand), Asif Noorani (Japan) Das Sharmila, Ghai Harjyoti, and Neerja Wable (India), Christine Malone and Kim South Hyde (South Africa), Pawel Ciacek (Poland), Andrei Ackles (Canada), Toni Parra (Spain), Inge Cootjans, Astrid DeJong, and Megumi Ishida (Netherlands), Lars Andersen and Julie Hoffmann Jeppesen (Denmark), and Ola Mobolade, Janette Ponticello, Bill Brannon, Dave Hluska, Doreen Harmon, Wes Covalt, Ariana Marra, Brian LoCicero, Brian Gilgren, Dusty Byrd, Mark Karambelas, Heather Fitzgerald, and Christina Swatton, all from the United States. I would also like to thank Lightspeed Online Research in the U.S. and not to forget Knots sample provider in Japan,

Since writing my last book, *BRANDchild™*, I've had the good fortune to meet people across the globe who have inspired me and helped to make this book more challenging and stimulating. I would like to thank Anne Pace (U.S.), Marie Louise Munter (Denmark), Kay Hannaford, Will, Warren, Bill, and Diane Greckers (Australia).

Thanks to Vibeke Hansen, Yun Mi Antorini, and Signe Jonasson, all from Denmark, and Anne-Marie Kovacs (U.S.), Henrik Kielland (Denmark), Warren Menteith (Australia), and the team at Google, who have all contributed individually with valuable information for this book.

Finally, I remain indebted to the many readers who sent me their invaluable feedback, and the thousands of people all over the world who answered endless questionnaires and patiently discussed the five senses in the many focus groups. Branding is about feelings, and their feelings have been an essential contribution to this book.

# Always updated at DualBook.com

I would love to claim that I had the idea to write about this topic . . . but in all honesty I didn't. The inspiration came from thousands of readers of *BRANDchild*™, who through my newsletter *BRANDstorm* and the book's online component, DualBook™, left me without any doubt that sensory branding had to be my next point of focus.

In keeping with my previous books, *BRAND sense* lives as a DualBook between the covers of this volume and online at www.DualBook.com. It's a pioneering and globally recognized concept, the first concept of its kind, allowing constant updates of this book as they occur. When you purchase a copy of *BRAND sense* you receive automatic membership in DualBook for one year.

As such there's no final page to *BRAND sense.* The subject continues online as I follow the evolving world of sensory branding. Every chapter remains open-ended, and when you go online you can experience the interactive dimension of the book.

## How to activate your FREE membership

Visit www.DualBook.com and click on the icon "Activate your membership" in the upper left corner of the screen. Your identification code is 90723834. Type this in and it will activate your personal membership. As soon as your membership is activated, you have free access not only to all the updates and other information related to *BRAND sense* but also to my two previous books, *BRANDchild*™ with Patricia B. Seybold, and *Clicks, Bricks & Brands* with Don Peppers and Martha Rogers, Ph.D.

*BRAND sense* is much more than this book. If you wish to explore

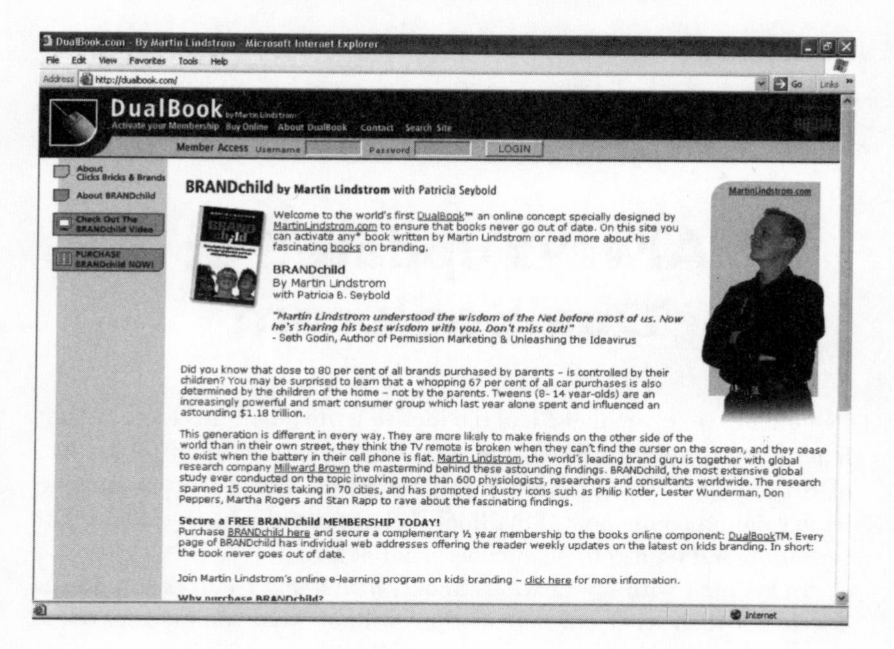


the world of sensory branding further, there will be the *BRAND sense* symposium, which will be covering 49 cities and 29 countries; the executive *BRAND sense* workshops; and the *BRAND sense* DVDs. Information about these are provided at DualBook.com, MartinLindstrom.com, and BRANDsense.com.

The concept of sensory branding combines channels in new and interesting ways, and I hope this gives you a stronger idea of how sensory branding will most likely change the way we create brands—forever!

# Index

**INDEX**

**INDEX**

# About the Author

Martin Lindstrom founded his own advertising agency at the age of twelve. Needless to say, his background is highly unusual. The rapid growth of his career has made him one of our time's most respected branding gurus, recognized by the Chartered Institute of Marketing.

He sits on several boards globally, and his clients include Disney, Pepsi, Philips, Mars, Mercedes-Benz, Kellogg's, Microsoft, and Reuters, to name a few. Lindstrom is a former BBDO executive, global chief operating officer of British Telecom/LookSmart, and founder and CEO of BBDO Interactive Europe and Asia Pacific.

His reputation for earning "a guaranteed standing ovation" at his compelling lectures on branding has accompanied his popularity as a consultant, columnist, and author. Lindstrom publishes his observations on branding in a weekly column, which reaches an audience of more than one million people in thirty countries. His books on branding, written with such industry icons as Don Peppers, Martha Rogers, and Patricia Seybold, are sold worldwide and have been translated into more than fifteen languages.

Over the course of twenty years of hands-on marketing experience, Lindstrom has conceived a revolutionary set of principles that transform marketing strategies into positive business results. He rejects the old rules of the industry that conceptualized branding as an art form composed of vague commercials and awareness messages. Instead, his unique vision is scientific and process-based. It makes branding the driver of sales and profits, and consequently the centerpiece of business.

Please visit the author online at MartinLindstrom.com or BRAND-sense.com.

# About the Author

# PRACTICAL MEDITATION

# PRACTICAL MEDITATION

## SPIRITUAL YOGA FOR THE MIND

B.K. JAYANTI

STERLING ETHOS
An imprint of Sterling Publishing Co., Inc.

New York / London
www.sterlingpublishing.com

**Library of Congress Cataloging-in-Publication Data**

Jayanti, B. K.
Practical meditation : spiritual yoga for the mind / B.K. Jayanti.
p. cm.
Originally published: Deerfield Beach, Fla. : Health Communications, c2000.
ISBN 978-1-4027-6626-8
1. Yoga, Raja. I. Title.
BL1238.56.R38J38 2009
294.5'436—dc22 2008045833

10 9 8 7 6 5 4 3 2 1

Published by Sterling Publishing Co., Inc.
387 Park Avenue South, New York, NY 10016

Distributed in Canada by Sterling Publishing
c/o Canadian Manda Group, 165 Dufferin Street
Toronto, Ontario, Canada M6K 3H6
Distributed in the United Kingdom by GMC Distribution Services
Castle Place, 166 High Street, Lewes, East Sussex, England BN7 1XU
Distributed in Australia by Capricorn Link (Australia) Pty. Ltd.
P.O. Box 704, Windsor, NSW 2756, Australia

Manufactured in China


Sterling ISBN 978-1-4027-6626-8

## CONTENTS

# FOREWORD

*Practical Meditation* is highly recommended for those who are beginning to discover the beauty and strength of their own inner world. More and more, we human beings are realizing the need to develop the power of the mind as we try to keep up with the ever-increasing pace of life and its complexities. The nine lessons in this book are based on the teachings of the Brahma Kumaris World Spiritual University, an international organization offering people of all backgrounds an opportunity to learn meditation and to deepen their understanding of themselves. The lessons can be practiced alone with great benefit, and there are a number of specific meditations to help the reader bring theory into practice. However, it is particularly beneficial to study the lessons while attending a free course in Raja Yoga meditation at any one of the Spiritual University's eight thousand centers worldwide. The atmosphere at these centers and the experience of being with other Raja Yoga meditation practitioners greatly enhances the understanding gained from the study of this book. By introducing this form of meditation into your own life, you not only gain real peace of mind, but also a more positive and healthy attitude.

SISTER JAYANTI
*European Director, Brahma Kumaris*

# 1 | WHY MEDITATE?

People want a variety of things from meditation. Some seek peace, others seek self-control, some wish for greater inner power, and some a sense of stillness. Of all the reasons to meditate, the ones which are most often expressed are peace or peace of mind. At first sight there doesn't seem to be much difference between the two, but on closer inspection, we find that they are asking for different things. Peace is simply an experience, whereas peace of mind is a way of life. At some time or other we have all enjoyed a moment's peace, however fleeting. To simply experience peace is actually not so difficult. Peace is something easily attainable through the practice of meditation because this is what meditation is specifically designed to give. However, to attain peace of mind implies that you want to experience peace constantly.

While you go about your daily life, you need to be in control of yourself to the extent that you can have whatever experience you choose when you choose to have it.





To experience constant peace of mind, you need something more than just a meditation technique. After all, in the middle of a dispute with the bus driver over change, you can't just sit yourself down and spend five minutes delving into the deep recesses of the self to regain that temporarily lost inner peace. It is the experience of peace, previously gained through meditation, which you need to be able to use later on in your practical life, especially at times when it is not easy to be peaceful. If you can't use meditation to bring benefit to your daily life, is it really of any use to you?

Therefore, the emphasis in this meditation course will be a double one: first, to teach a simple, effective method of meditation called Raja Yoga, and to discuss and experiment with ideas on how to deepen the experiences gained; second, to look at the reasons behind stress and tension in your life so that, through understanding, you can begin to change the root causes using the power gained through meditation; and also to clarify how you can translate peaceful feelings into peaceful actions, so that peace becomes peace of mind.





## WHAT IS MEDITATION?

Meditation is the process of getting to know yourself completely, both who you are *inside* and how you react to what is *outside*. Above all, meditation is enjoying yourself in the literal sense of the word. Through meditation, you discover a very different "me" from the person who may be stressed or troubled, the person who may seem superficially to be "me." You realize that your true nature, the real you, is actually very positive. You begin to discover an Ocean of Peace right on your own doorstep.

There is a lovely Indian story about a queen who lost her valuable pearl necklace. In great distress, she looked everywhere for it, and just when she was about to give up all hope of ever finding it, she stopped and realized it was right there around her own neck! Peace is very much like this. If you look for it outside in your physical surroundings or in other people, you will always be disappointed, but, if you learn where and how to look for peace within yourself, you will find that it has been there all the time.

The word *meditation* is used to describe a number of different uses of the mind, from contemplation and concentration to devotion and chanting. The word itself is probably derived from the same root as the Latin word *mederi*, meaning "to heal."

Meditation can certainly be looked on as a healing

process, both emotionally and mentally, and, to a certain extent, physically. The simplest definition of meditation is "the right use of the mind," or "positive thinking." Its purpose is not to deny thoughts, but to use them correctly.

Most forms of meditation employ two main practices: concentration exercises, often using an object such as a flower or a candle, or the repetition of a mantra. A mantra is a sacred phrase, word, or sound that is repeated constantly, either loudly, silently, or in thoughts only. It translates literally as *man* (mind) and *tra* (to free); so, a mantra is that which frees the mind.

Raja Yoga meditation does involve concentration, but no physical object is involved. The object of concentration is the *inner* self. Instead of repeating one word or phrase, as in a mantra, a flow of thoughts is encouraged, thus using the mind in a natural way. This positive flow of thoughts is based on an accurate understanding of the self, and so acts as a key to unlock the treasure-trove of peaceful experiences lying within.

## *Meditation Practice*

Sit in a comfortable position with the back straight. You can sit cross-legged on a cushion on the floor or, if this is uncomfortable, sit on a chair. Choose a quiet place away from noise or visual distraction. Gentle background music may be played, as this helps to create a relaxed, light atmosphere.

Position the book in front of you and, slowly and silently, read over the following words. Aim to experience and visualize the words in your mind so that you begin to feel what is being described.



### THOUGHTS FOR MEDITATION

*Let me imagine that nothing exists outside this room. . . .*

*I feel completely insulated from the outside world and free to explore my inner world. . . .*

*I turn all my attention inward, concentrating my thought energy on the center of the forehead. . . .*

*I feel a sense of detachment from my physical body and the physical surroundings. . . .*

*I become aware of the stillness around me and within me. . . .*

*A feeling of natural peacefulness begins to come over me. . . .*

*Waves of peace gently wash over me, removing any restlessness and tension from my mind. . . .*

*I concentrate on this feeling of deep peace . . . just peace . . .*

*I . . . am . . . peace. . . .*

*Peace is my true state of being. . . .*

*My mind becomes very calm and clear. . . .*

*I feel easy and content . . . having returned to my natural consciousness of peace. . . .*

*I sit for a while, enjoying this feeling of calmness and serenity. . . .*

Plan to practice repeating these or similar thoughts to yourself for about ten minutes at least twice a day. The best time is in the morning after a bath or shower, before you begin your day's activities. Another good time is in the evening, when your day's activities are over. During the day, while performing any activities, keep reminding yourself: *"Peace is my true nature."*

As you keep practicing this meditation, such positive and peaceful thoughts will enter the mind more and more easily, and peace of mind will become an increasingly natural state.

## 2 | WHO AM I?

The simple question of who you are seems easy to answer at first.

However, as soon as you start to think about it, you realize that giving your name or a description of your physical appearance does not describe the myriad thoughts, moods, actions, and reactions that compose yourself and your life.

Even a description of what you do becomes confusing, because every day you wear so many different hats. You may start the day as a wife or a husband. At work you may be a secretary, a clerk, or a teacher. At lunch you may meet a friend and in the evening an acquaintance. Which of these roles that you play is you?

In each role that you play, a different facet of your personality emerges. Sometimes you might feel that you have to play so many different and opposing roles that you no longer know what sort of person you are. When you meet your boss at a party or your parents and friends come to visit at the same time, you become confused as to how to behave. Not only have you fixed in yourself a special way of acting toward them, but in your mind you have also limited *them* to a certain role.

You are only able to relate to them as "your boss" or "your parents," not as simply other human beings. Yet you are quite aware that your true identity is not defined by the role you play. How can you think of yourself? Who are you really?

What is needed is something constant, safe, and stable. We get up in the morning and look in the mirror, and we seem to be much the same as we were yesterday. But we all know that this is an illusion, because gradually, over time, the body is declining; it is not stable or constant. In Raja Yoga, instead of adopting bodily identification, we start with our thoughts, awareness, or consciousness and identify with that, because our thoughts are always there, whatever age we are.

Their content may change, but our ability to think does not. First of all, you are a thinking, experiencing being.

Thoughts are not physical things that you can experience with the physical senses. You cannot see, taste, or touch a thought. Thoughts are not made up of matter or even brain cells. You are a nonphysical, spiritual being. The terms *self* or *soul* are used to describe this.

Your form is the only form that cannot be destroyed. It is something so small that it cannot be divided. It is something without any physical dimension. You, the soul, are a subtle star, a point source of light energy and consciousness. This subtle form is the source of everything that you do—all thoughts, all words, all actions. Whatever you do or say, it is you, the soul, who is performing that action through your body. The soul is like a driver and the body is the car. To be in complete control, the driver has to sit in the place where he has access to the controls and also can collect all the necessary information to make decisions. Each thought, leading to words and actions, begins with an impulse from the brain. In Raja Yoga, the soul is located in the center of the forehead near the brain. This knowledge provides you with a constant point of reference on which to focus your attention.



Your identity is a soul, and all the other identities—teacher, student, man, woman, father, mother, friend, relation, and so on—are simply different roles that you, the soul, play. Good actors can play any role. They will play their roles to the best of their ability, but will never actually think: "I am Hamlet" or "I am Cleopatra." They know that, however involved they are with their roles, at the end of the performance they will take off their costumes and resume their true identities. So, whatever role you, the soul, are required to play, you should understand that your true identity is a soul—a living, spiritual, eternal being. The body is simply your temporary physical costume.

The soul has innate peaceful, positive qualities. In meditation you can create an awareness of yourself as a soul.

This naturally leads to an experience of these peaceful, positive qualities. This is what is called *soul-consciousness*. It is not just something to experience while in meditation, but also as you perform action. As you become more aware of who is performing each action, you gain greater control over your thoughts, feelings, words, and actions. The natural consciousness of yourself as a peaceful being then fills all your actions, and the desire for peace of mind is fulfilled in a completely practical way.

In meditation you begin to think about your true identity.

You let thoughts about the soul and its qualities fill your mind. Initially it doesn't matter how fast the thoughts arise as long as they are moving in the right direction. If your thoughts wander away, gently bring them back again to peaceful thoughts of the self. As you become involved in the experience of such thoughts, they will gradually start to slow down, and soon you will be able to savor them. Just as when you are given something special to eat, you eat it slowly, appreciating each mouthful for its flavor and texture, so you begin to appreciate the experience contained in each positive thought. The simple phrase "I am a peaceful soul" comes to life as you begin to experience it.



This approach to meditation differs considerably from those that involve repeating a mantra or focusing on a candle or the rhythm of the breath. In Raja Yoga, sitting meditation is complemented by maintaining a peaceful soul-conscious state while performing everyday actions.

Through maintaining soul-consciousness in this way, you will continue to progress toward your aim of attaining constant peace of mind. A mantra is used purely for sitting meditation.

In Raja Yoga, you will bring your thoughts from

meditation directly into your daily life. This is the first and most important step in making meditation practical. As you go around doing things, you experience being a soul, acting a role through the body. Your consciousness becomes detached from your body. When you see another human being, you look beyond the name, body, race, culture, gender, and age and see, with the vision of equality, a soul like yourself who is simply playing a different role. This will help you to develop the qualities needed to remain peaceful all the time, such as tolerance, patience, and love.

Through understanding and experiencing your true qualities, you regain confidence and self-respect and are no longer pushed and pulled by the expectations of others. By remaining soul-conscious, you will stay in your true state of peace. This is something which does, of course, take time and patient effort to practice. The reward of practice is enjoyable in itself, and the greater benefits will accumulate over time.

## *Meditation Practice*

When you sit to meditate, choose the quietest place you can find, preferably in a room that you do not use very often.

If this isn't possible, sit where familiar objects won't distract you. Set this place aside purely for the purpose of meditation. This mental preparation will help your concentration. Start with ten or fifteen minutes. With experience, this time will gradually lengthen naturally. Soft or subdued lighting will help. A guided meditation CD can be used to guide the mind in a positive direction.

When you finish your meditation, just take a moment to reflect on what you have experienced; note how your mood has changed. This will emphasize your experiences and help you to appreciate what you are gaining through meditation.

One more suggestion that will benefit you greatly: Don't just meditate when you feel like it. The greatest progress is possible at the time when you really don't want to meditate or when you feel that you can't. That's the time when you need to meditate the most!

### THOUGHTS FOR MEDITATION

*I withdraw my attention from my physical limbs and senses. . . .*

*I focus on myself. . . .*

*I am listening through these ears. . . .*

*I am looking through these eyes. . . .*

*I am behind these eyes . . . in the center of the forehead . . . an eternal spark of life energy. . . .*

*This life energy empowers the body. . . .*

*I am a non-physical being . . . an eternal soul. . . .*

*I am the actor. . . . This body is simply my costume. . . .*

*I focus my thoughts on the point in the center of the forehead . . . a tiny point of conscious light. . . .*

*I feel completely detached from the body . . . peaceful and light. . . .*

*I am a star radiating light. . . .*

*I find deep peace and contentment within. . . .*

*I now know my true self . . . an eternal, pure, peaceful soul. . . .*

*I am in the Ocean of Peace. . . .*



*All conflict finishes. . . .*



*A deep, deep silence comes over me. . . . Om shanti.*

(*Om*, (I am), and *shanti,* (peace): "I am a peaceful soul.")

These thoughts are only a suggestion. Create your own similar thoughts if you prefer. Any thoughts based on the awareness of the self as a soul are valid. Think slowly and aim to experience each thought before moving on.

# 3 | SOUL-CONSCIOUSNESS

Why should the thought, "I am a peaceful soul" be any more beneficial to you than the thought, "I am a body"? In the previous lesson it was mentioned that this thought allows you to become detached from the role that you play. It is important to understand what is meant by this word *detached.*

It does not mean distant or inward-looking to the point of isolation, where there is a breakdown in communication. Nor does it mean that you become an uncaring observer of what is going on. It simply means to have the consciousness of being an actor. You play your part with great enthusiasm and love, but you do not let the expectations, burdens, and worries of outside situations or other people affect your own understanding of who you are: a peaceful being. In fact, the word most often used in conjunction with *detached* is *loving*.

By being aware of yourself as a soul, you can experience your natural qualities so that the feelings you associate with detachment are not of distaste or lack of concern, but of peace, love, and happiness.



How can soul-consciousness help you improve your attitude toward yourself and others? We often have the habit of comparing ourselves to others, seeing ourselves in the light of what we consider to be their merits or faults. This can sometimes lead to a feeling of hopelessness, self-criticism, and other equally negative states of mind. Through the experience of soul-consciousness, you will come to realize your own worth and stop comparing yourself with others. You will overcome self-criticism as you experience your true positive qualities and nature. This leads not to a feeling of superiority, but to a feeling of stability and self-confidence. Doubts in the self are replaced by a deeper faith in yourself.

If you understand that you are a peaceful soul, you will understand that the same is true of the people around you. Through this awareness, you will be able to relate to them on equal terms, that is, with what can be called the vision of brotherhood. Sometimes actions are totally opposed to this.

Someone may get angry with you and you feel threatened, replying sharply in return. Thus, a heated argument can develop. This is a reflection of body-consciousness. Instead of seeing the other as a peaceful soul playing a part, you see only the part and think it is the other's true nature.

If, instead, you have the determined thought to see others as souls, you will respond very differently to their anger. You will see their anger as something temporary and not intrinsic to their true natures. Instead of reacting angrily or being defensive, you will become detached or even giving. You can put yourself, quite naturally, into the position of being able to help them. You will recognize that their anger is a result of their own confusion. This positive attitude acts as protection; by thinking this way, you don't feel under attack. In addition, your stable, calm reaction will help to defuse the situation.

There are other ways in which soul-consciousness can help you to help others. You cannot give what you don't have yourself. When friends come to us in distress, often the most we can do is give sympathy. Although this is reassuring, it is not necessarily very helpful. When in trouble, what people need most is power and clarity. A bad situation has caused weakness and confusion in their minds, making it difficult for them to see things clearly. If your reaction and suggestions to them are not only sympathetic, but also filled with peace, power, and practicality, they can take with them not only comfort, but something which will be of positive value in helping them to solve their problems. For this, a strong and uncluttered mind is necessary.

Soul-consciousness also allows you to be natural in the company of others. This ease on your part helps them to relax, as they don't feel that you have expectations of them.

With soul-consciousness you aim only to see the good in others, to see not just apparent virtues, but hidden ones as well. This, in turn, helps others to realize their own positive virtues. A deepening love and respect for other souls will develop naturally as you recognize your spiritual kinship with them. You will realize that we are all part of one global family, sharing one world, one home.



### *Meditation Practice*

Meditation is being in the awareness of your natural qualities.

It is not a difficult thing. Nor is it something that you impose on yourself. You can't force yourself to meditate. In fact, the more hard effort you put in, the less likely you are to experience anything. Too much concentration will create a headache, and instead of refreshment and relaxation, there will be tension.

The first step is simply to relax. Many people

would consider being able to relax at will as quite an achievement in itself. In meditation, once you become relaxed, the worries and stress of everyday life dissolve and the mind is free to explore gentle themes. Your world is a creation of your own mind. This is why you fill your thoughts with soul-consciousness; thinking about peace helps you to experience peace.





The more relaxed you are, the deeper that pleasant, restful feeling becomes, until you have reached a powerful state of meditation, enjoying the quietness that is emerging from within. As you progress, meditation becomes much more than a relaxation technique. The object is not just relaxation; it is to become a peaceful person, to fill yourself completely with peace. The experience of peace, gained through simple relaxation, is a mere drop compared to the Ocean of Peace in which you can lose yourself through meditation.

Keep your thoughts or themes very simple in meditation; just two or three carefully chosen ones are enough. Repeat them gently, giving yourself plenty of time to explore the feelings behind them, for example, "Peace is just like sinking into a feather mattress . . . ," "Lightness is like floating on a cloud . . . ," "Love is a warm, golden glow inside your mind. . . ."

As you become more and more engrossed in such thoughts and experiences, you feel yourself gradually letting go of all worldly thoughts and tensions until you become light and free.





*I sit quietly, and can feel my physical body relax. . . .*

*I let tensions fall away, and I focus within, on my own inner being. . . .*

*I visualize my eternal identity . . . I am a point of light . . . a shining star, radiating light. . . .*

*As I focus on this awareness of who I am, I feel the connection with the body becoming one of detachment and love. . . .*

*I . . . the master . . . of this body . . . separate from this body. . . .*

*For a few moments, I allow my thoughts to go deep within this point of eternity. . . .*

*I . . . the eternal being . . . without a beginning, without an end . . . touch eternity . . . I . . . am . . . peace.*

Now I have become soul-conscious, aware of my true nature.

It is this lightness of consciousness that I want to bring into my everyday life, so that whatever problems and obstacles arise in front of me, I deal with them easily and effectively.



### THOUGHTS FOR MEDITATION

For the next few days just take up two or three simple themes or phrases, such as, "I am a peaceful soul," "I am a being of light and love, spreading these feelings to others and the world," or "I am a subtle point of consciousness, so different from the physical body."

Repeat these thoughts gently to yourself, allowing

them to sink in more and more deeply until your thoughts and your feelings match each other.

When this happens, the tension between what you think you should be doing and what you actually are doing disappears, and the soul feels content and full. In addition, practice seeing others as souls, seeing beyond the part to the actor who is playing the part.

Sitting down and experiencing peace is one thing; actually using it to transform your life is quite another. A great deal goes on between intention and action, and sometimes you might catch yourself saying, "I didn't want to do that, but . . ." or "Sorry, I didn't mean to say that." To be in full control of your life, you need not only to know, but also to understand the process through which an intention becomes an action.

For instance, a variety of raw materials

goes into the manufacture of a car: sheet metal, nuts and bolts, electrical wiring, paint, and so on. These raw materials can be compared to your experiences and intentions. As the raw materials pass through the manufacturing plant, they are processed and eventually emerge as cars. However, imagine that there is a consistent fault in the production process. You could repair each car as it comes off the production line, but this would be time-consuming and hard work. It would also be very frustrating, considering that the factory was not built to produce faulty cars.

Similarly, superficial changes to your actions will not result in profound life changes. Superficial change can work to a limited extent, but you will be continually faced with "faulty" actions from the "production line," and it will seem like very hard labor without much reward. Instead, you need to check the raw materials of your experience and also become familiar with the "production process" of your desires and actions. It is not enough for the engineer to know only approximately what happens on the factory floor. To repair a fault, he needs a detailed working knowledge of everything that is going on. The better he understands the machinery, the better he will be at identifying and repairing the fault. So, the more you understand about how you work, the easier it is for you to eliminate the actions that you don't want. Through meditation, you are checking the raw materials, ensuring that only the best quality is used and making sure that nothing is in short supply.



## INTENTION

So, what is the process of manufacture? The first and most obvious thing that comes between an intention and an action is a thought. Thoughts occur in the mind. In Raja Yoga, the mind is not seen as a physical thing but as a faculty of the soul, and therefore, nonphysical. Through the mind you imagine, think, and form ideas. This thought process is the basis of all your emotions, desires, and sensations. It is through this faculty that, in an instant, you can relive a past experience, produce happiness or sadness, or take yourself to the other side of the world.

When there is the thought, "I want a cup of tea," the relevant actions seem to follow automatically. However, is thought the only link between intention and action? What about the expression "Think before you speak"? Undoubtedly there must be thought before you open your mouth, or nothing would emerge; so, what is meant here?



There seem to be two aspects to thought. The first is the thought itself; the second is the awareness and understanding of that thought. It is the intellect that is used to understand the thoughts. Through this second faculty of the soul, you assess the value of what emerges in the mind. In the expression "Think before you speak," you are being asked to use your intellect and consider whether your thoughts are worth uttering. Some other functions of the intellect are reasoning, realization, discrimination, judgment, and the exercise of willpower.

Generally, you don't worry too much about what is going on in your mind. But now, for a few moments, you want to stop external activity and watch the internal activity: your thoughts.

*As I look within, I can see how my thoughts show awareness of the sounds outside . . . how they register memories of things of this morning, of yesterday. . . .*

*I see how they are filled with human images, and the impact of the people I've been with, the things I've heard, the feelings and moods of those around me . . . my mind has been influenced by all of these. . . .*

*For a moment, I take charge of my mind. . . . I create an image of a point of light . . . a thought of peace. . . .*

*I hold this thought, and as I continue to keep it in mind, it becomes more than a thought of peace. . . . It is a feeling of peace. . . .*

*A feeling that gives me comfort . . . and strength . . . and stays with me as I return to my everyday activities.*

## INTELLECT

The intellect is the most crucial faculty; through the intellect you exercise control over your mind and thus over yourself.

The purpose of meditation is to fill the intellect with power, thus making yourself clear-headed and perceptive, as well as developing firm resolve. The intellect is recognized by the effect that it has. For instance, someone explains something and you fail to understand it. So he tries explaining it in three or four different ways, but still you don't understand. Finally the fifth time, you *see the light*, that is, you realize what he means. This realization is the function of the intellect.

Another example might be the process through which you sort out a plan of action when faced with a choice of two or three possibilities. You weigh the advantages and disadvantages until your power of judgment tells you which plan is the most suitable. Like the mind, the intellect is a subtle nonphysical entity and belongs to the soul, not the body.

One of the most important realizations for the intellect to work toward comes in answer to the question, "Who am I?"

*I begin to be aware of my eternal form, my form of light. . . .*

*I, the spiritual being, with awareness, with recognition, am now the master. . . .*

*I am in charge of my physical body . . . and the master in charge of my own mind. . . .*

*I can feel the unlimited capacity of my mind . . . and as the master of my mind, I focus this unlimited energy onto the path of peace. . . .*

*I keep my thoughts focused on peace and truth . . . and my mind creates peace. . . .*

*I am that point of light . . . and I understand where my thoughts should go . . . the destination is peace . . . the path is peace. . . .*

*With the power of my mind channeled in this way, I radiate peace into the world. . . . I keep my mind in this one direction, of peace and truth.*

## SANSKARA

There is a third faculty of the soul, which comprises the impressions left on the soul by actions we have performed.

These impressions can be referred to by the Sanskrit word *sanskaras,* for which there is no simple translation. Habits, emotional tendencies, temperament, and personality traits are all built up by sanskaras imprinted on the soul through each action it has performed. Sanskaras create the personality in the same way that individual frames of a feature film make up a story. Every action is recorded, whether it is a physical movement, a word, or even a thought. As you live your life, you are making an imprint on the celluloid, the soul. All the thoughts that occur in the mind are due to the sanskaras. Personality, the most fundamental feature of each individual, unique soul, is determined by these sanskaras.

The mind, intellect, and sanskaras function together in a cyclic pattern that determines how you behave, what thoughts you have, and even what mood you are in. First, the mind produces thoughts, evidence that the intellect judges.

On the basis of that judgment, an action is performed or not performed. The action, or inaction, creates a sanskara that, in turn, becomes part of the evidence in the mind.

A good illustration of this is the formation of a habit such as smoking. The first time you are offered a cigarette, many thoughts, both for and against, arise in the mind: "It's bad for my health," "I wonder what it tastes like," "It is very easy to get addicted," "Everyone else does it" and so on. On the basis of these thoughts, the intellect makes a decision. Let's suppose that it makes the decision to try a cigarette. A sanskara is created by that action, and the next time you are offered a cigarette that previous action becomes part of the evidence in the mind, as a memory: "I smoked one before." If you decide to smoke one again, the repetition deepens that sanskara, just like planing a groove in a piece of wood, until eventually the evidence in the mind, urging you to smoke, has become so overwhelming that no evidence for not smoking remains.

The intellect has now become very weak, even defunct. There is no longer a choice or judgment to make. There is just the strong thought rising in the mind: "Have a cigarette!", and you perform the action automatically. You are no longer in control; your past actions in the form of sanskaras are ruling your present.





However, you can also use this mechanism to create peaceful, positive sanskaras. As you sit in meditation, you will experience yourself as a peaceful soul. This experience forms a sanskara. The next time you are about to get angry, through force of habit, the mind will present contrary evidence: "I am a peaceful soul." This forces the intellect to make a decision.

As the intellect gains strength of will through meditation, it becomes easier to act on peaceful sanskaras, as opposed to negative sanskaras. Thus the intellect begins to control both the mind and actions. You, the soul, become the master of the present. You are no longer the slave of your past. Gradually you will reach a position where you choose to put into action only those thoughts that will lead you to experience permanent happiness and contentment.

### *Meditation Practice*

Take one aspect of yourself that you want to change. A few times a day, create just one or two very powerful positive thoughts that will help change that negative habit or character trait. Do this with all the energy and enthusiasm you can muster. This will create a very powerful sanskara.

When that positive thought for change comes into your mind again, it will bring with it the experience of enthusiasm. This will help you to put that intention into action at the appropriate time. For example, if you want to give up the habit of criticizing people, throughout the day keep creating the positive thought: "I see all as peaceful souls; instead of criticizing their weaknesses, I will only see their virtues and specialities" or "I must first change my own weaknesses before criticizing the weaknesses of others."

## 5 | KEEPING THE BALANCE

To continue to progress toward your aim of attaining constant peace of mind, the most important thing is balance. If a car is too heavily weighted on one side, the driver will find it difficult to control and maneuver. Problems will arise with the tires, suspension, and so on. The same can happen to you if you pay too much attention to sitting in meditation, being introverted, and not enough attention to relating peacefully to others. You may become withdrawn, living in your own inner world, instead of the "real world" outside. You may find that relationships with others become difficult.

There are four aspects to bear in mind to avoid such an imbalance. If equal weight is given to all four, you will remain balanced while making natural, easy progress.

These four aspects are: *knowing, being, becoming, and giving.*





## KNOWING

*Knowing* refers to the understanding of knowledge.

You have been given the basic facts: You are a soul; your true nature is peaceful; you have a mind, intellect, and sanskaras. Now you have to fit them together. These facts are like the pieces of a puzzle; it is only when they are fitted together in the correct way that the picture emerges. Each piece has a little bit of a pattern on it; on its own, it can only hint at what the complete picture is. By turning the information over in the mind, playing with it, matching it up to your life as it unfolds, you begin to create a coherent view. Once there is understanding, you begin to feel that you are in control of your situation. When there is understanding, your intellect remains clear, and you are able to act in a positive and effective manner. Knowledge allows you to be detached from potentially stressful situations.

*What is it that I now know? I know that my hands and feet, my arms and legs, are simply my body parts . . . they are not me. . . .*

*Even the whole of this body is not me . . . it is my instrument, or vehicle. . . .*

*But who am I, the one in charge of this instrument? I now understand how I, the pinpoint of energy, the point of light within, am the one who uses this instrument. . . .*

*I, the soul, look out through these eyes. . . . I, the soul, receive the information coming in through these eyes. . . . I, the soul, decide to communicate, and so I use this mouth. . . . I am the one who decides what it is that I wish to communicate. . . .*

*I, the being of light, have the capacity to decide what information I pick up through these ears. . . .*

*I am the master of this physical instrument. . . . I am in charge. . . .*

*I know . . . who . . . I am.*

## BEING

*Being* refers to yoga, the experience of meditation.

Even if you can sort out all the logical connections between the bits of information you have received, unless you have a grasp of their true meaning, you cannot really say you have understood them. For instance, you could learn some simple phrases in Hungarian and be able to repeat them in the correct order, but unless your teacher had explained the meanings of the words, the phrases would be of absolutely no use to you.

So, how are you to understand what words like *peace, love, soul,* and *detachment* mean? You understand these concepts only by experiencing them. The experience of peace makes peace a reality. It also gives you a basis of trust and faith, for it is when the concept and the experience coincide that the soul can feel secure. Practical experience of the theoretical knowledge which you have been given verifies the knowledge.

This leads to trust in the knowledge; through that trust and sense of truth you build a stable foundation.



*Stepping back . . . inside my own skin . . . coming to the awareness of the being that I am . . . I explore my original state of being. . . .*

*Within my being, in my original state, there is cleanliness . . . purity. . . .*

*I accumulated dust when I traveled, but when I began my journey, I was clean . . . pure. . . . As the dust is removed, and that pure, clean state shines through. . . .*

*I can feel peace. . . .*

*Peace is my natural state of being . . . this is who I am. . . .*

*In this state of purity . . . and peace . . . I rediscover the love that is within . . . a love that is altruistic . . . love for myself . . . love for each member of my human family . . . love for the Supreme. . . .*

*Purity, peace, love, and joy . . . these are my natural qualities, my true state of being.*





## BECOMING

*Becoming* refers to your actions.

In the last paragraph, harmony between knowledge and experience was emphasized. If there is any contradiction, trust and stability disappear. What is vital here is harmony between what happens internally and what happens externally. To sit in meditation and experience yourself as a peaceful soul, and then immediately afterward to become angry with someone, renders that peaceful experience meaningless, and the soul feels lost and confused. Meditation must be made practical; its positive power must be reflected in action. You will actually become that which you experience in meditation.

Putting the results of meditation into practice must, on the whole, be a conscious thing. It won't happen miraculously, without your paying attention. It is easy to see why this is true if we again consider how the soul performs actions through the cycle of mind, intellect, and sanskaras. Even though you are creating peaceful sanskaras in meditation, the old peaceless sanskaras will continue to create negative thoughts in your mind, sometimes very powerfully. It is only through conscious choice within the intellect that you can discriminate and change your behavior.

What is important to understand here is that you will never experience progress unless you make an effort to change your negative actions and habits. However good your experiences in meditation are, if they are constantly contradicted by your actions, you will continue to create negative thoughts about yourself. Your mind will become a battlefield instead of a haven of peace.

*Let me take a journey back to my original state of being, in which I was pure, clean, without a blemish, without a stain. . . .*

*In that state of cleanliness there was comfort . . . there was peace. . . .*

*Through my journey, I accumulated a heap of rubbish . . . a huge amount of dust. . . .*

*And now, I clean out that which doesn't belong to me . . . I become that which I was, that which I am, that which is my true nature. . . .*

*I connect with my own original form . . . and I act in that consciousness . . . I become that, here and now. . . .*

*Purity, peace and love are within me . . . I let these qualities emerge. . .*

*I become clear and clean again.*

## GIVING

*Giving* refers to harmonious and altruistic relationships with others.

Although becoming peaceful automatically helps your relationships with others, you still have to pay attention to this area, mainly because it is your relationships with others that spark off peacelessness within you. It is easy to be friendly and giving when those around you are also friendly and giving. Unfortunately, in today's world, we often find ourselves in interpersonal conflicts, ranging from mildly uncomfortable to openly hostile. In these situations, the practice of giving is your protection. It protects you from experiencing negativity, but also benefits the other soul who is unfortunate enough to be feeling aggressive. You cannot give and receive at the same time, so, thinking only peace and good wishes means there is no room for responses of fear or resentment or the awakening of anger within you.

These types of situations are the tests that face you every day. How you cope in these instances is the true measure of your progress. When there is victory, you realize that you truly understand some aspect of knowledge. If, however, you do become angry or get careless, the desire to get it right next time sends you back to review the knowledge for deeper understanding. Sometimes you will be in a position to help others directly by sharing your own positive experiences. When this happens, having put things in your own words will make you realize how much you have understood. Every time you return to the knowledge, you will have moved a little bit further forward. So, natural progress is taking place.





Giving should be done without the desire for return or reward. It should be a natural process, simply motivated by the wish to share with others positive experiences you have internalized. Feeling happy and content is the natural reward of your positive actions. Without desires and expectations, your giving becomes truly altruistic. When you have practiced meditation for some time, giving becomes something beyond words.

The knowledge and meditation experiences will

become so much a part of you that simply by being your true positive self, you will give the experience of peace and virtuousness to others.

*In the awareness of my original state of being, my original treasures of purity, peace, and love emerge fully. . . .*

*And in a very natural way, whatever I am, whatever I have . . . I transmit to the world. . . .*

*In this awareness of my eternal treasures, I send out good feelings . . . pure feelings . . . toward each member of my human family. . . .*

*I radiate purity . . . I am a being of peace. . . .*

*What do I have that I can share with the world? . . . It is peace. . . .*

*My thoughts . . . my vibrations . . . of peace . . . spread out into the universe. . . .*

*I contribute to the creation of a world of peace by sharing thoughts of peace. . . .*

*I am a being of love . . . and I feel this warm glow within me . . . a warmth that comforts and supports and empowers others. . . .*

*It is not the love of possessiveness or attachment . . . the love that binds me to one or two . . . it is the love that is inclusive, that connects me with the whole world. . . .*

*I give . . . pure love . . . to my world family.*





When all these four aspects of *knowing, being, becoming,* and *giving* are in harmonious balance, the soul will be at peace with itself and in harmony with others. This state of practical soul-consciousness has been termed "*jeevan mukti*" or "freedom in life."

### *Meditation Practice*

Slow down! Give yourself time to think before you act. Give your new peaceful sanskaras a chance to be put into practice.

Give yourself permission to have the time to practice.

Concentrate during the time you have available.

Short periods of regular meditation will increase the benefit you experience. Naturally, over time, the periods you remain in meditation will lengthen and the benefit will continue to increase.

Practice being detached from your own thoughts. Create the thought: "I, the soul, am in the cinema, watching my thoughts come up on the screen of my mind." As you watch your thoughts, they will begin to slow down. Sort through them, separating the positive and the negative or wasteful (mundane) ones. Replay the best thoughts, and allow your thoughts to lead you into the experience that lies behind them.

If you find that your mind is still too active or at all negative, first concentrate on that basic thought, "I am a peaceful soul."

Observe the direction in which your thoughts flow from that positive source. Sometimes the mind will naturally go in a positive direction when you sit for meditation, but at other times it needs to be firmly steered and guided to avoid crashing into the rocks of negative emotions and thoughts!

## 6 | KARMA

While you may have associated yourself completely with your physical body, you may not have realized that every action had such a deep impact on you. Now, with the recognition of the self as a soul, you should become aware that every single action leaves an imprint, a record, which you carry with yourself eternally.

Up until now, we human beings have found it very difficult to classify exactly what is right and what is wrong.

Throughout history our definition of

right and wrong has been changing. Different cultures and religions have come up with different definitions and classifications. Even within the same religion, people of different generations have different ideas of right and wrong. Even if you don't consider the external situation at all, but look only within yourself, you will find that your understanding fluctuates a great deal. In childhood, your understanding was on one level; in adolescence it changed; in maturity it has changed yet again.





As you are influenced by the atmosphere or the words of human beings, your intellect wavers in its own judgment. So, can you possibly arrive at a point where you know absolutely what is right and what is wrong? Not while you are limited by this physical costume. The religion into which you were born, as well as the limitations of gender, of age, and of culture, will all color your ideas, thoughts, and judgments.

By maintaining the consciousness of your true identity—a peaceful soul—you will be able to accurately understand what is right and wrong. This is simply because, in soul-consciousness, the soul can only experience peace, happiness, and love. So, it can only perform actions based on these qualities. These actions will be beneficial actions, bringing happiness and positive results. In body-consciousness there is not the pure intention behind action. Our actions are performed with selfish ulterior motives, such as greed, ego, and possessiveness, and are therefore nonbeneficial actions which give sorrow and bring negative results. It is the consciousness with which we perform action that is important.

The law of karma, of action and reaction, is applicable to the spiritual sphere and is absolute. It states: "For every action there will be an equal and opposite reaction." *Opposite,* of course, means opposite in direction. Whatever interactions you have with others, you receive the equivalent in return. This means that, if you have given happiness, you will receive happiness in return, and if you have given sorrow, you will receive sorrow in return. The law is simple, and when understood in its full depth, it can give insight into the significance of events in your own world and in the world at large. In Christianity this law has been understood by the saying, "As you sow, so shall you reap." It is also known as the law of cause and effect.

Understanding this, when you see certain effects, you will now realize that effects can only take place if there is a cause. So karma (action) is the cause, and the fruit of karma is the effect. Generally, when you see the fruit of your karma, you might tend to forget that you are responsible for these effects. If the fruit of karma is bitter rather than sweet, you might point the finger of blame at others and say that others are responsible for your suffering. If there is the effect of sorrow, you now should understand that you have been responsible for the cause of sorrow. Understanding the law of karma makes you take total responsibility for your own situation, your state of mind, and indeed your whole life.

Sometimes only half the law of karma is understood, that is, the part concerning destiny. Someone may think helplessly: "Whatever is happening to me now is because of my past actions; so, there is nothing I can do about it. It is my fate."

When there is understanding of the law of karma and awareness that you are responsible for your own situation, you will develop tolerance, acceptance, and endurance, qualities that may have been missing before. However, more important, the other side of the law of karma teaches that, if you now perform pure, beneficial actions, you can create your own positive future in the direction of your choice. Not only are you not a slave to destiny, but the understanding of karma philosophy makes you the creator or master of your own destiny. You may even be able to inspire others to create a positive destiny for themselves through the example of your own beneficial actions.

Any negativity of the past has led you into karmic debts with those around you. Where you have in the past given sorrow, you must now repay that debt by giving happiness. You have to settle your past karmic accounts. However, just because you change your attitude doesn't necessarily mean others will change theirs. If you have courage and continue to give good wishes and perform pure actions in relation to other souls, gradually the karmic debts with others toward whom you have acted negatively will be repaid. Then you can be free from the bondage of karma.

The power to sustain this effort of settling past karma can come through meditation or yoga. As you come to understand your own true nature more fully, you can understand that this is the true nature of everyone. You can see through the mask of negativity and relate to the soul directly. This will help you not to create further negative karma; you will not react badly to the negativity of others.





*I . . . the soul . . . have been performing karma through this physical costume of mine. . . . I carry the imprints of this karma, good and bad. . . . I have been seeing the results of my karma externally, in the world around me. . . .*

*And now, I come to the awareness of my eternal state . . . a state in which I have no karmic bondage. . . .*

*This is the state I want to attain once more . . . the state of freedom . . . the state of having settled all my karmic debts, and of having created for myself a stock of good karma. . . .*

*In this awareness of I, the eternal being, I look back at the past and see the entangled threads of karma. . . .*

*Where did a connection begin? Where did it finish? The threads are so intertwined . . . it's difficult to differentiate. . . .*

*The web of karma has been one in which there has been a lot of pain . . . a few flashes of light, of hope, of joy . . . but the negativity of my karma created situations in which there was much suffering. . . .*





*Now, at this moment, in the presence of the Divine, I let go of my karma of the past. . . .*

*With God's love, I surrender my past karma to God . . . and I take light and might from the Supreme . . . so that my karma today will create a future filled with light. . . .*

*God's light shows me the path of righteous karma. . . . God's might . . . God's power . . . gives me the courage to reject negative karma and even mundane karma. . . .*

*I simply do that which is elevated and noble . . . so that my present is filled with light and might . . . and a future of happiness and love is assured.*

With soul-consciousness you will naturally give love and respect to others, and you will, in time, receive love and respect in return. Every action performed in soul-consciousness is an action through which you receive benefit, and thus it will benefit others. Karma begins in the mind as thoughts, the seeds of action. As is the thought, so is the result. Thoughts, like actions, spread vibrations and influence the surrounding atmosphere. Karmically there will be a return of those vibrations. Pure, peaceful, happy thoughts are the most valuable treasures in life. If you keep such beneficial thoughts in your consciousness wherever you go, you will create a pure atmosphere of peace and happiness from which others will greatly benefit.

Understanding the consequences of actions means you take care to do everything properly. Having little control over your actions is a sure sign that you have little control over your mind. This links up with the last lesson; if you slow down, you give yourself more time to do things properly. If something is done well, most likely it will not cause problems in the future. Jobs done in a rush often contain mistakes that have to be put right later, thereby causing more work. A good job done well leaves you with a peaceful mind. A careless piece of work pulls your attention back to itself again and again. If you concentrate completely on what you are doing in the present, this allows you to be in full control of both mind and body. You keep performing actions in soul-consciousness so that, no matter how much you have to do physically, you can remain light and peaceful.



### *Meditation Practice*

Divide your time for sitting in meditation into three segments.

Take three meditation themes or positive qualities that follow on from each other, such as stillness, silence, and power or lightness, peace, and contentment.

Consider each theme separately, creating thoughts that will lead you into the experience of each quality. Make sure that you have achieved an experience of the first quality before moving on to the second, and the second before moving on to the third. In this way you can gently lead yourself into deeper experiences in meditation, and you can begin to enjoy a variety of positive feelings and qualities.

# 7 | THE SUPREME

Throughout history, we human beings have sought many things. Above all we have each desired two things: happiness and a perfect relationship. If we achieve either one or both of these things, it becomes a constant struggle to keep them, and they usually prove to be temporary. If we want to achieve them on a permanent basis, we must look beyond the limited gains of possessions, money, and fragile human relationships.

Raja Yoga has two meanings: *Sovereign Yoga,* the yoga through which you can become the sovereign, the master of yourself; and the *Supreme Union,* or "union with the Supreme." This second aspect of Raja Yoga involves developing a relationship with the Supreme, the source of perfection, God as we perceive Him. Within this yoga or union you can fulfill your wish for inner happiness and your desire for a perfect relationship.

It does not even require that you first believe in God. It is useful to simply have openness to the idea that there may be a greater source of spiritual energy than yourself. Through your own experimentation in meditation, you can develop an understanding of this concept. If someone asked you, "Do you believe in the existence of Mr. X?" you would be inclined to want to meet him first before committing yourself. Under the circumstances, you would keep an open mind. Similarly, with the concept of a supreme spiritual energy, until there is direct experience, it would be unwise to commit yourself. Yet, if you want to have contact with the Supreme Being, there are certain things that you must know. First, you should know the form of the Supreme so that you will be able to have accurate recognition. Second, you need to know what "language" to use so that there can be communication. Third, you need to know where your meeting can take place.

In Raja Yoga, just as we have a very precise notion of the form of the soul, so we also have a very precise notion of the form of the Supreme. In fact, the Supreme is recognized as the Supreme Soul. So, He has a form identical to that of the human soul, that is, a point source of consciousness, a spark of light energy.

When we use the word *He,* this is not to imply that we think of the Supreme as male. The soul itself has no gender; it is only the body that has gender. Whereas the human soul takes a body, the Supreme Soul never has a body of His own, and so, is neither male nor female.

God never takes human birth. So, He never forgets his original qualities as we do. He remains eternally peaceful, blissful, and powerful. Our experience of our own original qualities is limited.

God is eternally the unlimited ocean of virtues.

He is completely full, and so never needs anything. This means He is totally benevolent, ever-giving. In fact, within Raja Yoga we have a particular name for the Supreme, and that is Shiva Baba. *Shiva* means benevolent. He is the only being who is truly altruistic, whereas we humans normally look for something in return, even if it is only the pleasure of giving.





God gives without any desire or expectation of return. *Baba* is a sweet and familiar name for Father; thus, Shiva Baba is the benevolent father of all souls.

The reason *Father* is used is because of the concept of a father giving an inheritance to his children. In this case, the inheritance received is of peace, love, knowledge, and happiness. The Supreme is also the Mother, the Friend, and the Beloved. In fact, whatever relationship or positive role you wish to see in Him, you can, because He is the unlimited source of all qualities, both male and female. So, whatever the situation, you always have a source of help and strength to draw on, a source that is only a thought away.

How can you communicate with this being? Meditation is about experiencing yourself, experiencing your own qualities. You create peaceful thoughts in order to experience peace. Paradoxically, the more you absorb yourself in that peace, the fewer thoughts you need. The communication with the Supreme is on this level. You come to know him through the experiences that you have of his qualities. You begin to feel those qualities surround you. My communication with God is primarily through silent experience. In deep silence you can lose yourself in the Ocean of Peace. With this experience you feel refreshed. You begin to understand your own qualities and specialties more deeply, and this brings confidence. You take power, which enables you to maintain a peaceful stage while going about your daily life.

*In the awareness of my eternal identity . . . I, the soul, become aware of an eternal connection . . . not only with all the human souls around me, but also with a being who is the Supreme. . . .*

*A soul . . . a being of light . . . with a form that is infinitesimal . . . and yet from that point of light, a radiance of infinite peace, love, joy. . . .*

*In the awareness of my own original state of peace . . . I can tune in . . . and connect with the Ocean of Peace. . . .*

*Waves of peace from the Ocean of Peace reach me and surround me . . . peace filled with sweetness . . . peace filled with strength. . . .*





*I, the child of the Supreme, realize that this is my eternal Parent. . . . This is the one who gives constant love, support, protection. . . .*

*This is my Parent who constantly cares for me, sustains me and guides me . . . the Ocean of Unlimited Love. . . .*

*This is the Being who is the Bestower, the Absolute, able to give constantly, and so needing no return. . . .*

*One who is benevolent, the Supreme Benefactor. . . .*

*Through my thoughts, I stay connected with the Supreme . . . I fill myself from the Ocean . . . I reclaim my original nature of peace, love, and joy. . . .*

You also need to know where to find Him. When you sit in meditation and go deep within yourself, a feeling of stillness comes over you. In that silence, your experience is that you are in an unchanging world, a timeless world. Yet this physical world is ever-changing. If your consciousness is tied to the physical, you can never get away from the passage of time. It is as though you have taken your consciousness beyond this world to another world. We call this place the soul world, the original home of the soul. It is a timeless world of silence and stillness, full of peace and power, a world of infinite golden-red light. This is also the home of the Supreme Soul. Taking yourself there, you begin to experience His unlimited qualities of peace, love, purity, bliss, and power surrounding you. Through this most perfect of all relationships, you take power and guidance so that you can clear your karmic debts of the past and create a peaceful, happy, and stable future.

*As I connect with One, the magnetism and power of the Supreme lift my consciousness beyond the physical dimension . . . into a world of light. . . .*

*I find myself in a place of infinity . . . a region where there are no borders . . . a place of stillness . . . of silence . . . of perfect purity. . . .*



*This is my home . . . a place where I feel so comfortable, so at peace . . . I am with my Supreme Parent . . . I belong to this Mother and Father . . . to this home. . . .*

*I experience my original state of stillness, of purity, in my home. . . .*

*This is a place of rest. . . .*

*Down below, the world stage is a place of action, and I will journey back there in a few moments. . . .*

*But for now, I can be up here in my home . . . with my Parent . . . learning to be the observer . . . learning to be free. . . .*

## *Meditation Practice*

Sitting in a quiet and relaxed atmosphere, slowly read over the following thoughts about the relationships one can have with the Supreme.

### THOUGHTS FOR MEDITATION

*When I meet the Supreme in the land beyond sound and movement, the only things that exist are the feelings in my heart. . . .*

*It is my open heart that God reads . . . He knows what it is I truly desire . . . He fulfils that need. . . .*

*What is required of me is honesty, purity, and clarity in my own mind to enjoy fully this meeting with my Supreme Father—Baba. . . . Baba fulfils my deepest needs and wishes in many ways through many relationships. . . .*

*As the Father, God gives me love and understanding. . . . As His child, I have the spiritual birthright to his inheritance, the unlimited treasure-store of all his perfect virtues and powers. . . .*

*I also experience the sweetness and comfort of God as the Mother, in whose lap the soul can rest in tenderness and care. . . .*

*I can share all my thoughts and hopes with Baba, my Friend, and even my doubts and problems, for there is nothing to hide from a true friend. . . . I can enjoy a heart-to-heart conversation at any time I wish, in any place. . . .*





*As the Teacher, God fills me with truth. . . . He has an answer for every question, advice for every need, revealing to the self all the secrets of time and eternity, unraveling the mysteries of creation, so that the meaning of life becomes so clear . . . for in God I have found the perfect Teacher, the source of truth . . . and He is also my Liberator and Guide, freeing the self from all sorrows and suffering . . . guiding me along the path to freedom and happiness. . . .*

*As my one Beloved, God is the comforter of my heart. . . . With God as my Beloved, the search for true love ends and the experience of contentment and completeness begins.*

*Having the experience of all these relationships with God provides me with everything I need . . . fulfilling all my pure desires and dreams. . . . The seed of all these relationships is love. . . .*

*Behind every thought and action of His is pure love and the wish only to bring benefit to the soul . . . to uplift . . . to purify me. . . .*

*God's love is unlimited and endless.*

Sir Isaac Newton described a clockwork world, ticking away inside a cosmic clock. Time was an absolute thing. A second was a second—no longer, no shorter. Wherever we are, in our living room or on a pulsar millions of light-years away, whatever we are doing, it ticks away, independent of any outside influences. This is a common-sense view of time, and if we thought like this, we would be thinking along the same lines as Western scientists did for nearly three hundred years after Newton.

Then, at the beginning of this century, an enigmatic figure called Einstein presented a theory, the implications of which rocked the foundations of three centuries of work. This was the theory of relativity. Part of what Einstein said was that the only way one can measure time is by clocks, be they water clocks that drip every second or mechanical clocks that tick. Basically all clocks move, and, therefore, time is dependent on movement; time is not totally independent.



He wasn't the first person to think like this; he was, however, the first person to formulate a usable mathematical theory about it (containing that famous equation, $E = mc^2$). To understand this clearly, let's return to Newton's theories.

What Newton said was that there is a cosmic clock out there ticking away, against which one can measure things. This clock of Newton's isn't a "real" clock, but it's like a solid idea.

Time is passing by, independent of whatever is happening.

Now let's do a little thought experiment. Imagine

that we all go to sleep one night, and when we wake in the morning, everything is moving at half speed. According to Newton, "real" time is still ticking away up there, and, in fact, we are taking twice as long to do everything. In the final reckoning, a nine-to-five job has not taken eight hours, but sixteen hours.

Einstein says that we have no way of knowing, when we wake up on the morning after this strange event, that everything is working at half its previous speed. There is no real time there, by which to measure everything that happens. Time is a measure of our activities. If all the timepieces in the world have slowed down to half speed, then time itself has slowed down to half speed. In other words, Newton's time is rigid and Einstein's time is elastic.

When we woke up that morning, everything would have seemed normal. We wouldn't have felt that everything had slowed down, because the only way to judge that would have been by comparing it with something else which had not slowed down. In other words, things would still be moving at the same speed relative to each other. How can these two ideas help you in a practical sense? First, according to how you regard time, you are either its master or its servant. With a Newtonian outlook, you become its servant, as "time waits for no one," and so you feel that you have to rush around, cramming as many activities as you can into every relentless second. Now, let us consider how it is more helpful to have Einstein's view of time.





In this case, time is dependent on the rate of change. What is the changing thing within yourself whose speed is going to govern how fast time appears to be moving? It is your own thoughts. If you slow down your thoughts, time will appear to expand. If you speed them up, time contracts. It's not that you slow down your thoughts in the same way you would slow down a record; you simply leave space between each thought or even between each word. You then become aware, not only of the thoughts, but also of the free spaces.

Awareness of these peaceful spaces between your

thoughts brings you right into the present and gives you the feeling that there is room to maneuver, time to spare.

When you first approach a new activity—for instance, the first time you cook from a particular recipe—you read each instruction carefully. Then, thinking only about that, you perform the relevant action, returning to the cookbook again only after you have completed it. This mode of operation makes sure that everything is done correctly and the best possible result is attained. Having given yourself time and space to do the job well, on completion of the task, you feel satisfied.

Compare this with the situation where the instructions, in the form of thoughts, follow each other in rapid succession, not waiting for each instruction to be put into action before the next one arises. The result is that, as you are doing one thing, your mind is badgering you to get on and do the next thing. You feel under pressure. You feel that you do not have enough time to do all the actions correctly.

Consequently, upon completion, you often find that the job has not been done well. Instead of a feeling of satisfaction, there is stress and tension.

So, it is not just the speed of thoughts that is important, but the speed of thoughts relative to actions. If the speed of your thoughts (your instructions to yourself) matches the speed with which you can do things, you will remain free from stress and tension and you will feel that there is time to do things properly. The effect is that you feel as though you are *creating* time for yourself.





Another advantage that immediately becomes apparent when you slow your thoughts is the ease with which actions and reactions can be controlled. Great sportsmen have control over their minds when practicing their sport. This is clearly visible with athletes; so precise and clear are the instructions given to the body that each step taken or shot made seems effortless and totally economical. There is no wasted physical effort.

The spaces we leave when we slow down our thoughts allow us to change direction easily and immediately. When thoughts race, they gather momentum like a car going at full speed. If we are required to make an unexpected left turn, we must slam on the brakes, upsetting ourselves and the people in the cars behind us. We will probably overshoot the turn and have to waste time and effort, reversing and finally making the correct maneuver.

When this happens in our minds, the emergency stop leaves us shaken and confused, and it can be disturbing to those around us. However, the spaces between thoughts are like times when we are temporarily stationary. From a stationary position we can move in any direction we choose, smoothly and easily, without causing discomfort to anyone.

This practice of slowing down thoughts and giving ourselves more time is helpful in many ways. Above all, it allows us to be soul-conscious much more easily. Those spaces give us time to enjoy sweet feelings of peace and contentment, which are natural qualities of the soul.

Undoubtedly, Newtonian time is the sort of time that governs the physical world around us. Without this solid framework to refer to, things would be most disturbing. However, through Einstein's "window," you can escape this physical world and fly to that timeless expanse, the soul world. This highest dimension is timeless, as there is no movement, just constant stillness. In your spiritual home you can learn how to slow down your thoughts to such an extent that they stop altogether, and now, in total silence, stillness, and contentment, you discover the beauty of eternity.





### *Meditation Practice*

Practice the habit of saying "Past is past." Keep facing forward.

If something negative happens, don't feel guilty about it. Simply have the determined thought to conquer it.

Rechannel the energy that usually goes into guilt or regret into positive thought and willpower, so that the soul says, "Yes, I am making efforts to change and improve myself."

## THOUGHTS FOR MEDITATION

*I experience myself as a bodiless being . . . a source of light, peace and power in a world of light. . . .*

*I feel that everything is totally still . . . timeless . . . Nothing is changing. . . .*

*I experience deep contentment. . . .*

*There is nothing more that the soul desires. . . .*

*I am with God in my eternal home of silence. . . .*

*I absorb the Ocean of Peace. . . .*

*I remain with the unlimited source of all virtues and fill myself totally. . . .*

*I am overflowing with light and peace . . . spreading the qualities of the Supreme into this world.*

# 9 | THE EIGHT POWERS

Eight powers are specifically developed through Raja Yoga. At first they might seem like eight qualities, but the word "power" has been used, in this context, for a particular reason. What is the difference between a quality and a power? A quality is something that others can sometimes see in us, but it is also sometimes hidden. It is something which others appreciate, but do not necessarily feel that they can possess. A power is something that cannot remain hidden; it is a constant source of inspiration to others so that they too can change themselves and become powerful.

The eight powers are:

1. *Introversion*, or the power to turn within;
2. The power to *let go* of thoughts of the past that are no longer useful;
3. The power to *tolerate*;
4. The power to *accommodate*, or adjust to situations and people;
5. The power to *discern*, or discriminate between true and false;
6. The power to *judge* priorities and make decisions;
7. The power to *face* obstacles or losses;
8. The power to *cooperate* with others.



It is important not just to know what these powers are, but even more important, to understand when and how to use them. For instance, if you are constantly tolerating someone's bad behavior and the situation is getting steadily worse, perhaps you should really be using the power to face them; that is, to have the courage to say, in a peaceful but firm way, that such behavior is not acceptable. The eight powers are such that in any situation, there will be at least one power that is appropriate for your use. The correct choice of that power will depend on your remaining calm and having a clear picture of the situation in question.





1. **The power to turn within** is the foundation of all the powers. It brings the strength to remain peaceful and positive while in the midst of life's challenges. Naturally, your thoughts must be concentrated and engaged when you perform an activity, but during any task you can also turn your thoughts within and practice returning to your state of inner peace. In this way, your thoughts do not continue to be involved needlessly, and you waste no mental effort.

This is the true power of control, which brings great strength to the soul.

*The wind, the rain, the sun, the storms . . . all of this is the nature of the outside world . . . but just as a tortoise turns inside its shell, I turn inward to the essence of my own being . . .*

*I connect with the point of peace that I am. I find that inner balance, that equilibrium, that strength within. . . .*

*Turning inward, I connect with truth, with reality. I understand who I am and what I must do. . . .*



*With my awareness of truth and reality renewed, I step out into the world, now able to perform action with the consciousness of truth.*

2. **The power to let go** of wasteful (mundane) thoughts in soul-consciousness means that you can travel light, packing only that which is necessary. Not carrying around negative and wasteful thoughts keeps you free from both mental and physical tiredness.

This economy brings power and a completely positive outlook.

*I step out of the cycle of the past, present, and future and look at the cycle of time as an observer. . . .*

*At this one moment, I can see the past . . . I can see the present . . . I also see the future. . . .*

*I pack up the things of the past, having learned the lessons that are useful . . . I put away the things that are not necessary, that are no longer required. . . .*

*I stay with the essence, with truth, with reality, so that my present thoughts are elevated . . . and my actions will be likewise. . . .*

*When I look into the future there is nothing but light . . . only light . . . and goodness . . . I learn to love this power of packing up.*



3. **The power to tolerate** difficulties involves the ability to go beyond the influence of negative situations, to be able to not react, even in thoughts. If someone offers you insults, criticism, or anger, or if there is physical suffering, with the power to tolerate you can remain peaceful and happy. On the basis of soul-consciousness, you will be able to give love, like the fruit tree which, when pelted with sticks and stones, gives its fruit in return.

*Connecting with my own inner treasures . . . connecting with the divine source, the Supreme . . . I am filled with the fruits of all attainments. . . .*

*I have all that I need . . . I am overflowing . . . I have enough to be able to share generously with others. . . .*

*Faced with aggression, I don't respond in kind . . . I respond by sharing the fruits of my attainment. . . .*





*I give peace where there has been aggression . . . love, where there has been animosity . . . compassion, where there has been insult. . . .*

*A tree laden with fruit will share only fruit, nothing else . . . I receive the fruit of all attainments from God, and I share these with the world.*

4. **The power to accommodate** is the power to be above any clash of personality or nature, to be able to mold and adjust yourself as the situation requires. You should not create conflict in any situation. Just as an ocean can accommodate all the rivers flowing into it, so you should not reject anyone or anything.

Instead, you are able to change relationships and circumstances through the power of good wishes.

*Connecting with the Supreme, I am drawn from the limited into the unlimited. . . .*

*My heart stretches and grows . . . I receive love and wisdom from the Supreme . . . and I am able to be generous. . . .*

*As an ocean is big enough to accept everything that flows into it, I too, connected with the Supreme, am able to absorb, to accept, to accommodate. . . .*

*I can adjust to different situations, and personalities, and all the challenges the world brings me. . . .*

*Keeping myself connected with the Unlimited, my capacity to love becomes unlimited.*

5. **The power to discern** is the ability to give correct value to the thoughts, words, and actions of yourself and others. Just as the jeweler can discriminate between false and real diamonds, so you should be able to keep positive, worthwhile thoughts and

discard negative, harmful ones. It is the negative thoughts that often cloud true discrimination, and you will eliminate these through meditation.

*I become detached from my physical body, stable in the awareness of myself, the soul . . . and God's light touches the intellect, and makes it divine. . . .*

*The dust accumulated on the intellect falls away . . . and the divine intellect sparkles, pure and clean, absolutely free. . . .*



*Connected with the Supreme, I understand all the different ingredients of a situation . . . all the different facets of the diamond, as well as the flaws . . . the variety of energies in relationships. . . .*

*This understanding brings light . . . and I can see clearly what is false, and what is beautiful and true.*

6. **The power to judge** allows you to make clear, quick, accurate, and unbiased decisions. For this, you need to be above the influence of situations and

the emotions and opinions of others. You also need a clear understanding of what is right and wrong. Raja Yoga meditation provides this strength and clarity of the intellect through greater self-understanding and a detached perspective.

*My connection with the Supreme creates clarity. . . .*

*I understand all the components of a situation . . . I have no bias toward this person or that person. . . .*

*My heart is full, so my judgment is not distorted by desires. . . .*

*I can decide between priorities . . . I know what needs to be done . . . the correct path for me to take is clearly lit in front of me.*

7. **The power to face** obstacles in life (courage) is developed by meditation, through which you experience your original nature of peace and become detached from the consciousness of the physical costume. You are then able to observe and see beyond

problems and difficulties, to discover a positive side to something that seems totally negative; this gives you the strength to face these obstacles.

*Connecting with the Almighty Authority, as the child with the Parent . . . I receive as my inheritance all the powers of the Supreme. . . .*

*I have been transformed and uplifted, so that I, the soul, am the master almighty authority. . . .*





*I have strength and courage . . . I do not fear the situations that come. . . .*

*God's power gives me the capacity to face whatever comes my way . . . I am able to remain loyal and true to the path of truth and peace. . . .*

*Facing all adversity, I remain loyal to my Supreme Companion . . . with the power of his company, I overcome all obstacles.*

8. **The power to cooperate** with others requires the vision of soul-consciousness, with which you can see all

as your brothers and sisters. This attitude of brotherly vision creates unity and strength within a group. This power of cooperation will make any task seem easy.

*God's gifts of all the different powers have been preparing me for this . . . God's strength and light make it easy for me to give and receive cooperation. . . .*

*Having shed my ego and weakness, I see the special qualities of each one in my family . . . I'm able to work with them, cooperate with them. . . .*

*Coming together with the awareness of the one Supreme, we all give our share of cooperation . . . and the task is accomplished . . . a mountain is lifted. . . .*

*"I," "my," and "mine" have all melted away . . . only the Father, and the Father's task, remains . . . and with this one thought, we come together . . . to serve.*

These eight powers become more and more effective in your life as you become expert in applying them to situations, as required.

# IN CONCLUSION

This small book provides a brief introduction to the concepts underlying the study and practice of Raja Yoga meditation.

After reading this book, it is important to clarify how you can use this information in a personally meaningful and satisfying way. The aim of Raja Yoga is to provide a means by which you can become the master of your own mind and your own destiny, and thus acquire constant peace of mind. So, it is essential for you to proceed at your own pace in a way which feels comfortable. Above all else, you need to develop faith in yourself.

Having faith in yourself means having the courage to explore what you are capable of experiencing. Blind faith is not a reliable foundation. If one places trust in things that are not understood, sooner or later that trust is going to be shattered.

First, there must be understanding; for this, knowledge is needed. In Raja Yoga very specific knowledge is given. Think about this knowledge, understand it, and develop a clear picture in your mind. Think about the implications of being a pure, peaceful soul. What effect will it have on your practical life? How will it affect your relationships with others? Does the law of karma provide adequate explanations of current situations in your life and in the world? If God exists, will it be possible to have a relationship with Him?

Only if you consider the implications of the knowledge will you have some sort of measure against which to place the practical experience of meditation and daily interactions.

When you sit in meditation, you can see if your experience matches up to the information you have been given. If you practice soul-consciousness during the day, you can see if this brings the results predicted by the knowledge. There will then be a basis for faith; it will rest on the firm foundation of your own experience, not only in meditation, but also in practical life.

Faith needs an aim. To measure your progress, you need to know where you have come from and where you want to go.

Knowing where you have come from is not difficult; knowing where you want to go is more subtle. Knowledge tells you, "I am a peaceful soul" or "Om shanti," but how do you translate this into experience? Deep inside you desire peace, but it is like a half-forgotten memory. That desire for peace is being prompted by a sanskara, a sanskara telling you that you must have had an experience of very deep peace before. You are not aiming for something you have never known before. Your aim is simply to rediscover that forgotten feeling of being so peaceful that you are ever-content. When you experience peace in meditation, it feels so natural and easy. It is an effortless thing. Your aim is to be in that experience constantly.

Whatever you are doing, whoever you are speaking

to, whatever is happening around you, you remain "Om shanti," lost in the Ocean of Peace, spreading vibrations of peace to others. Faith needs both understanding and experience to sustain it. So, simply make sure that you combine ongoing study of this knowledge with the practical experience of meditation. In this way your life becomes an inspiring source of positivity and happiness for yourself and for those with whom you interact.

## ABOUT THE AUTHOR

Sister Jayanti is a spiritual teacher and leader, a pioneer and an emissary for peace. She has a vision and experience that is truly global and deeply spiritual. Perhaps this is because, among other factors, she is a child of two cultures. Born in India of Sindhi parents who migrated to England when she was eight years old, she is a blend of Eastern wisdom and Western education and culture. At the age of nineteen she began her life's path of spiritual study and service with the Brahma Kumaris World Spiritual University. At the age of twenty-one, she dedicated her life to making a difference in the world. She has trained for more than thirty years with some of the world's most remarkable yogis. As a result, she herself is an extraordinary meditator and teacher and has developed a clarity and purity of mind that is exceptional.

Sister Jayanti is also a much sought-after speaker around the world. Her natural wisdom and gentle, though powerful, personality have touched and inspired millions of people throughout the world. She is the European Director of the Brahma Kumaris World Spiritual University and assists in coordinating the university's activities in more than ninety countries. She is also its representative to the United Nations in Geneva.

## ABOUT THE BRAHMA KUMARIS WORLD SPIRITUAL UNIVERSITY

http://www.bkwsu.org

**International Headquarters**
P.O. Box No. 2,
Mount Abu 307501
Rajasthan, India
Tel: (+91) 2974-38261 to 68
Fax: (+91) 2974-38952
Email: abu@bkindia.com

**International Coordinating Office & Regional Office for Europe and the Middle East**
Global Co-operation House
65-69 Pound Lane
London NW10 NHH, UK
Tel: (+44) 208 727 3350
Fax: (+44) 208 727 3351
Email: london@bkwsu.org

### *Regional Offices*

**Africa**
Global Museum for a Better World
Maua Close, off Parklands Road, Westlands
P.O. Box 123, Sarit Center
Nairobi, Kenya
Tel: (+254) 20-374 3572
Fax: (+254) 20-374 2885
Email: nairobi@bkwsu.org

**Australia and Southeast Asia**
78 Alt Street
Sydney, NSW 2131, Australia
Tel: (+61) 2 9716 7066
Fax: (+61) 2 9716 7795
Email: ashfield@au.bkwsu.org

**The Americas and the Caribbean**
Global Harmony House
46 S. Middle Neck Rd.
Great Neck, NY 11021, USA
Tel: (+1) 516 773 0971
Fax: (+1) 516 773 0976
Email: newyork@bkwsu.org

**Russia, CIS, and the Baltic Countries**
2 Gospitalnaya Ploschad, Build. 1
Moscow – 111020, Russia
Tel: (+7) 499-263 02 47
Fax: (+7) 499-261 32 24
Email: Moscow@bkwsu.org

**Brahma Kumaris Publications**
www.bkpublications.com
enquiries@bkpublications.com